3D Printing of Metal

3D Printing of Metal

Editor

Zhanyong Zhao

MDPI • Basel • Beijing • Wuhan • Barcelona • Belgrade • Manchester • Tokyo • Cluj • Tianjin

Editor
Zhanyong Zhao
School of Materials Science
and Engineering
North University of China
Taiyuan
China

Editorial Office
MDPI
St. Alban-Anlage 66
4052 Basel, Switzerland

This is a reprint of articles from the Special Issue published online in the open access journal *Metals* (ISSN 2075-4701) (available at: www.mdpi.com/journal/metals/special_issues/3D_metal).

For citation purposes, cite each article independently as indicated on the article page online and as indicated below:

LastName, A.A.; LastName, B.B.; LastName, C.C. Article Title. *Journal Name* **Year**, *Volume Number*, Page Range.

ISBN 978-3-0365-6865-2 (Hbk)
ISBN 978-3-0365-6864-5 (PDF)

Contents

metals

Review

Current Status and Challenges of Powder Bed Fusion-Based Metal Additive Manufacturing: Literature Review

Naol Dessalegn Dejene * and **Hirpa G. Lemu**

Department of Mechanical and Structural Engineering and Materials Science, Faculty of Science and Technology, University of Stavanger, N-4036 Stavanger, Norway
* Correspondence: nd.dejene@stud.uis.no

Abstract: Powder bed fusion (PBF) is recognized as one of the most common additive manufacturing technologies because of its attractive capability of fabricating complex geometries using many possible materials. However, the quality and reliability of parts produced by this technology are observed to be crucial aspects. In addition, the challenges of PBF-produced parts are hot issues among stakeholders because parts are still insufficient to meet the strict requirements of high-tech industries. This paper discusses the present state of the art in PBF and technological challenges, with a focus on selective laser melting (SLM). The review work focuses mainly on articles that emphasize the status and challenges of PBF metal-based AM, and the study is primarily limited to open-access sources, with special attention given to the process parameters and flaws as a determining factor for printed part quality and reliability. Moreover, the common defects due to an unstrained process parameter of SLM and those needed to monitor and sustain the quality and reliability of components are encompassed. From this review work, it has been observed that there are several factors, such as laser parameters, powder characteristics, material properties of powder and the printing chamber environments, that affect the SLM printing process and the mechanical properties of printed parts. It is also concluded that the SLM process is not only expensive and slow compared with conventional manufacturing processes, but it also suffers from key drawbacks, such as its reliability and quality in terms of dimensional accuracy, mechanical strength and surface roughness.

Keywords: additive manufacturing; selective laser melting; defects; process parameter; powder bed fusion; surface roughness; porosity; microstructure; residual stress

Citation: Dejene, N.D.; Lemu, H.G. Current Status and Challenges of Powder Bed Fusion-Based Metal Additive Manufacturing: Literature Review. *Metals* **2023**, *13*, 424. https://doi.org/10.3390/met13020424

Academic Editor: Zhanyong Zhao

Received: 22 January 2023
Revised: 11 February 2023
Accepted: 16 February 2023
Published: 18 February 2023

1. Introduction

A new class of technology known as additive manufacturing (AM), commonly referred to as three-dimensional (3D) printing, is concerned with the direct production of physical items from computer-aided design (CAD) models utilizing a layered manufacturing process. A revolutionary change from craft to mass production occurred in the manufacturing paradigm prior. AM's arrival resulted in a fresh paradigm shift away from mass manufacturing and toward mass customization [1]. AM has grown exponentially because of its crucial role in the fourth industrial revolution [2]. Due to that, AM has become a key technology for fabricating customized products owing to its ability to create complex objects with advanced attributes (new materials, shapes) [3], near net shape, and buy-to-fly ratio [4,5]. According to the consensus among researchers and the industrial community's findings, metals based on AM technologies have tremendous application potential that cannot be matched by traditional manufacturing technologies [6], and metal AM is a promising technology in the next generation of the manufacturing process [7]. Moreover, the AM market, though small compared to traditional manufacturing, has grown rapidly, which is shown in Figure 1 of Ref. [1]. According to Koutiri et al. [8], since 2003, there has been a 42% increase in parts produced by AM. Wohler's report in the 2017–2018 reporting period [1] also reported that the overall 3D printing industry grew by 21%. Moreover, according to the Wohler's association in 2020, an industrial insight report on the AM market, the average annual growth of the AM market was predicted to be over 20% in the next five years [9].

The ISO/ASTM 529000:2015 standard has categorized the AM methods into seven main groups, including binder jetting, directed energy deposition, material extrusion, powder bed fusion, sheet lamination and vat photopolymerization [10–12]. PBF method is widely employed in a variety of industrial areas, including the aerospace, energy, transportation [13,14], biomedical, automobile and jewelry industries [8,15–17]. PBF is the focus of this paper, and details are given in Section 3.

Quality and reliability are major concerns in state-of-the-art Industry 4.0 technologies, including AM [18]. AM technologies have recently gained more attention due to their ability to manufacture complex parts, so the quality and reliability of AM is an emerging area. The challenges of PBF-produced parts have become a hot issue among stakeholders. Tapia et al. [6] demonstrated a consensus among experts and stakeholders in the aerospace, healthcare, and automotive fields that metallic AM parts are still not sufficient to meet their stringent requirements. For successful adoption, it is necessary to have a comprehensive understanding of the materials ecosystem in AM. This paper discusses the present state of the PBF process and technological challenges, with a focus on SLM. Decisively escaped discussing all types of PBF processes, instead focused on SLM and slightly stating electron beam powder bed fusion (EB-PBF).

2. Materials and Methods

In this review work, the literature coverage considered is comprised of open scientific sources published mainly in the last 10 years. To obtain sufficient coverage of the works reported on the topic of the PBF-based AM process, we used the following scientific databases to search through the literature: scientific.net, Scopus, Elsevier, Science Direct, web of science, Google scholar, springer, semantic scholar, PubMed, JSTOR, DOAJ, MPDI, research gate, SAGE, BASE, SCISPACE, Taylor & Francis and others. The search keywords used in English language were additive manufacturing, powder bed fusion, selective laser melting, defects, process parameters, quality, surface roughness, porosity, microstructure, residual stress, etc. Other descriptive terms were also used in some cases to be specific with the objective of the review.

3. Powder Bed Fusion—Based Metal Additive Manufacturing Process

PBF is a group of AM technologies where an energy source is used to selectively bind or melt powder particles to build parts layer by layer to the desired geometry [15]. The first PBF process, selective laser sintering (SLS), was developed at the University of Texas [2,12,19,20]. According to Ahmadi et al. [21], SLM began in 1995 at the Fraunhofer Institute ILT in Aachen, Germany.

All other PBF processes significantly change the latter basic approach in one or more ways to improve machine productivity. The basic schematic representation of the PBF process is illustrated in Figure 1. PBF is widely applied and possibly the most evolved AM technology available [22]. According to Vafadar et al. [23], 54% of the metal AM market belongs to PBF processes as of 2020. Korpela et al. [24] stated PBF-process is the most evolved metal AM technology to produce engineering components. The PBF machine manufacturer has their own specific commercial names, such as SLS, direct metal laser sintering (DMLS) and SLM, which is also known as laser powder bed fusion (L-PBF) or electron beam melting (EBM) [13,23–27]. Although some systems have the word "sintering" in their names, the majority of metal PBF systems in use today completely melt the particles rather than sintering them [24,28].

Comparing the effectiveness of those PBF and other processes is currently becoming a research concern. For example, the PBF shares common parameters with directed energy deposition (DED). The PBF system utilizes a powder deposition method that consists of a recoating mechanism that spreads a powder layer onto a substrate plate and a powder reservoir. Once the powder is evenly distributed, PBF uses thermal energy from a heat source (laser or an electron beam) to trace the geometry of an individual layer of slices from a 3D model on the powder bed's surface [29]. DED can use the same source of energy and feedstock as PBF. However, DED deposits a melted material feedstock through a nozzle onto a surface [30]. The similarity between DED and PBF under the same source of energy becomes an interesting concern. DED has a high deposition rate and is preferable for larger component manufacturing due to the slower process of PBF [31]. PBF methods have an advantage over DED methods in producing parts with higher resolution and quality [30]. Babuska et al.'s [7] findings strengthen the aforementioned statements, where the mechanical performance of Fe-Co components is produced by PBF and DED under a laser beam. The part L-PBF was characterized as high strength (500–550 MPa) and high ductility (35%), while L-DED results in low strength (200–300 MPa) and low ductility (0–2.7%). Even though there are fundamental manufacturing approaches, the process parameters involved at different levels are not explained in the study.

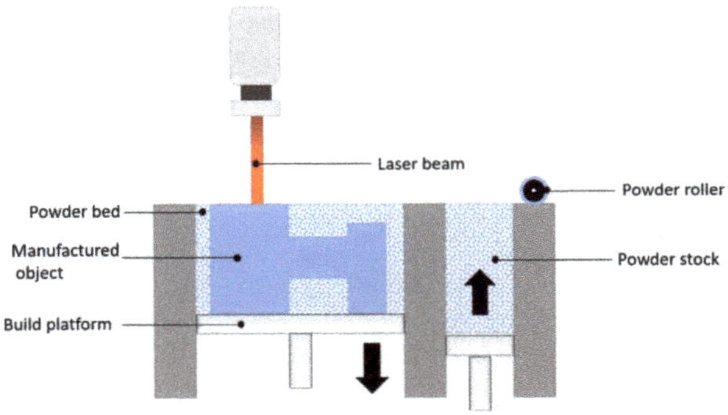

Figure 1. Schematic representation of PBF (Reused from [32] with permission in accordance with Creative Common CC BY license, open access).

The comparison among the PBF sub-categories is also another concern of the continuous improvements in R&D. The printing process of both L-PBF and EB-PBF is nearly the same [4] or a similar process [33]. However, there are several differences that need consideration as they affect the quality of the final components. Basically, EB-PBF was introduced to overcome the challenges of L-PBF, such as power and speed [15]. EB-PBF uses a high-energy electron beam 3000 W [15] or (~3500 W) [34]) and proceeds in a vacuum to control the oxidation of highly reactive metals like titanium and magnesium [19,20]. The presence of a vacuum atmosphere in EB-PBF favors slower cooling rates [14], which leads to lower residual stress [4,14,35]. However, EB-PBF suffers from poorer dimensional accuracy [29,36] and surface roughness [29,37] and is limited to only conductive metals [4,19]. Thus, it is comparatively less applied to the current technology in the industry. In contrast, L-PBF has a relatively broader application due to its capacity to produce fine-size features (i.e., to small pool size) and intricate features [4]. L-PBF uses lower power (100–1000 W) [14,38], relatively good dimensional accuracy and roughness [14,39]. The challenges of L-PBF are high porosity due to the presence of inert gas or nitrogen [40], fast cooling rate that leads to high residual stress [14] and a slow scanning speed [19] compared to EB-PBF. However, L-PBF (SLM) emerges as a leading candidate to manufacture mission-critical applications [41–43], such as aerospace and defense applications [44]. Therefore, the focus of this study is principally on the SLM, which is currently the most used metal AM technique for fabricating parts.

4. Process Parameters in Powder Bed Fusion-Based Metal Additive Manufacturing

As mentioned earlier, PBF is possibly the most evolved AM technology. However, the production process is still expensive and slow compared to conventional manufacturing. Moreover, the components are typically semi-finished and need post-processing. According to Ladani et al. [15], the most interesting aspect of PBF-AM is the possibility of the potential control of final products through a bottom-up approach; i.e., microstructure, flaws and other properties can be controlled through the process by intelligently varying the parameters. Several authors agree on the primary contribution of process parameters to the formation of defects. Hence, monitoring the root causes of defects and controlling them provides important insights into enhancing the quality of final products. Numerous researchers stated the influential process parameters for SLM as laser input energy, powder material and scanning speed [45]. Murugan et al. [16] considered the layer thickness, scan speed, hatch spacing, size of the powder particles and orientation of the layer. Javidrad et al. [46] used scanning speed, laser power and hatching space to determine the effect of volume energy density (VED) on the microstructure of Inconel 625. From the process mapping, it is identified that the thermal gradients, melt pool size and residual stress are affected by the powder feed process [33,47].

As Ladani et al. [15] stated, the outcome of SLM largely depends on process parameters and the interplay of physical phenomena. The great desire of the manufacturing industry is to be more profitable through high production rates, low wastage, no defects and less effort. So, the significant contribution of process parameters to the final product is revealed via various studies. However, the parameters guaranteeing the highest production speeds are not disclosed with the best achievable accuracies. For example, layer thickness increases have a significant positive impact on building time and a negative impact on accuracy [24,40]. This implies that as the production rate increases, the desired quality decreases, so several process parameters need to be considered in SLM. However, some of those can be predicted in advance, while others are challenging to predict because they are generated through rigorous processes. Therefore, bottom-up monitoring and controlling process parameters are vital techniques. This review categorizes the aspects of the bottom-up process parameters sources under powder, laser, powder material, and printing chamber environment parameters, as shown in Figure 2.

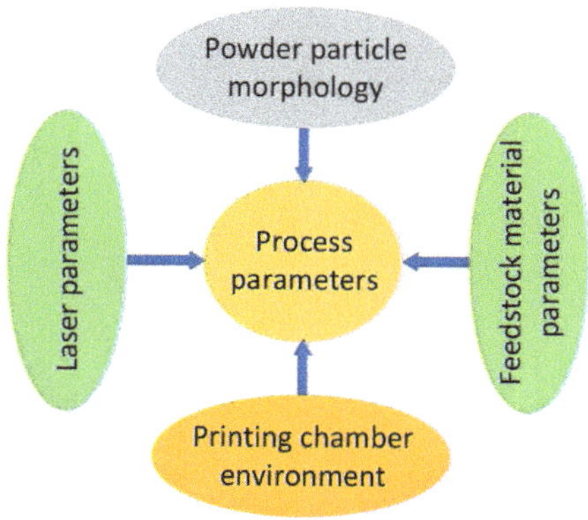

Figure 2. The source of process parameters for PBF-AM.

4.1. Aspects of the Powder Particle Morphology

The PBF-based metal AM received attention from industries and R&D due to its rewards, high degree of customization and the minimal buy-to-fly ratio [5]. The hard materials that would be difficult by other manufacturing processes are some of the reasons why SLM tends to be selected. Moreover, SLM can produce complex feature parts, and powder can be reused. However, the SLM process came with additional properties of powder particle size distribution (PSD), shape, density, flowability and other powder morphology, which have considerable impacts on the qualities of components. Those parameters can be categorized as the single particle (shape, smoothness, PSD, impurities, composition, moisture, etc.) and bulk properties (apparent or bulk density, tap density, Hausner ratio, flowability and physical properties) [48]. The parameters related to the powder morphology are the pre-defined or pre-process parameters. The powder morphologies, such as PSD and shape characteristics of individual particles, such as sphericity and uniformity, have significant effects on the bulk properties of powder as well as the final part. This is due to their critical impact on determining the optical penetration depth [49,50].

The single-particle morphology determines the conductivity due to the voids and level of connection among particles. This affects the penetration depth and leads to melting pool size and shape, which are responsible for the variability in part properties and several defects [49]. The bulk properties of powder particles, such as flowability, have various interrelated effects on the process [48]; a very small change in the PSD influences the flowability, which has been noted to impact packing density [51]. The flowability is reduced due to the satellite particles (<10–15 μm) connected to large particles impacting the packing density due to increased mechanical interlocking powder flows in SLM [40]. However, the fine grain size enables achieving better layer thickness, which improves surface roughness. According to Karapatis et al. [52], experimental results show the apparent density of thin powder layers increases from 53% to 63% of solid material when adding 30 vol % of fine powder to the coarse one. The mechanical performance improvements in steel components in the SLM process are due to refined microstructure by small particles [43]. This indicates the contribution of particle size in microstructure is also noticeable.

The material properties of components are extremely sensitive to pre-process parameters [53–55]. Spierings et al. [56] explored how three different powder granulations affect the mechanical characteristics, surface quality and part density of the produced SLM steel. The finding indicates the coarser particles result in a rougher surface and lower density. In the other case, fine particles are easily melted and favor good part density and surface quality, whereas large particles have the benefits of ductility, mechanical strength, hardness and toughness [57]. The flowability, which is the effect of both fine and large particles, impacts layer-to-layer variation in parts, and as the powder size increases, the yield strength and ultimate tensile strength (UTS) increase [49,51,58]. This reinforces that PSD has a concern with repeatability in the material properties of produced parts. Ladani et al. [15] mentioned that powder feedstock quality has a direct effect on the quality of fabricated parts and the repeatability of the process. Therefore, the PSD and other morphologies of powder have various effects on the parts, such as repeatability.

The powder particle source of challenges can be from the powder production techniques to the part manufacturing. However, there is inadequate coverage in the literature on the powder production techniques' improvement with the alignment of the PBF process. Powders prepared with different methods also showed different mechanical properties on the components due to the difference in their microstructure [15]. Samples fabricated using gas atomized (GA) powders showed higher microhardness than the samples prepared using mechanically alloyed plasma spheroidized (MAPS) powders with almost similar composition [15,59]. Sanaei et al. [60] find feedstock material as the first source of defects based on the atomization process and whether it carries entrapped gas. In addition, there are claims that the overestimation of the SLM material utilization is 100%. It has been reported that a 95% material efficiency is more reasonable [61]. So, it is crucial to realize how laser heat affects recycled metal powder and in what way that processing affects the final parts' microstructural and mechanical properties [62]. The impurities and powder recycling are keen aspects of monitoring the qualities of the part as bottom-up controlling. The leftover powder from each build is recycled, then returned to the system for further use. However, the powder can be exposed to air during handling, and hence, the formation of surface moisture and oxides is possible [62]. In addition, as one major aspect, the environment of the SLM processing chamber is very important to fabricate oxide-free parts.

Despite using protective inert environments and a shielding inert gas flow to limit the oxygen content in the working chamber, there is always a chance of a small percentage of unwanted oxygen content (~0.1–0.2%) present during the SLM process [63]. Oxygen existing in the surrounding atmosphere entrapped inside the powder particles causes various defects like porosity, void formation and mechanical properties [16,40,57]. Moreover, considering the percentage of the recycled, as well as the number of recycling times, additional unseen characteristics of parts will be revealed. Lewis [5] addressed the effects of powder reuse PSD, the aspect ratio of particle morphology and flowability. However, the number of recycling is not addressed as there is a high chance of repetition in the SLM to be cost-effective.

The comparison of sieved recycled and fresh powder of AlSi10Mg microstructure was conducted, and slightly similar results were obtained [62–64]. However, the experimental work considered a limited number of recycled powders. The production of high-quality and fully dense components generally depends heavily on the grain shape, PSD, surface morphology and overall purity of metallic and alloying elements.

4.2. *Aspects of the Laser Process Parameters*

In addition to the powder particle morphology, the SLM parts are sensitive to printing process parameters (e.g., beam size, power, scan rate and others) [54,65]. The SLM shares common parameters with seam welding [25,66], so most of the descriptions are taken analogically in different studies. Therefore, the defects that occur due to unrestrained process parameters in micro welding, such as keyholes that can be avoided by expanding the beam size [67], are expected in SLM.

The interaction of the laser and the powder is a complex phenomenon due to multiple phase changes and variable absorbance [27,49,57,68]. That is due to the powder morphology and the rapid timescale interaction with the focused nature of a laser spot on microscopic spatial scales. Even when operating on these small scales, minor perturbations can have a significant effect [49]. Therefore, the factors affecting energy absorption and quality of components are the beam spot size, the shape of the deposit geometry and protective gases [27]; this is due to their impacts on the size and shape of the melt pool [33,54]. Mudge et al. [69] indicated that the laser spot size needs to increase with powder feed to increase the deposition rate, which increases production volume [42]. However, doing that would lead to a coarse surface finish due to the increase in melt pool size. The size and shape of the melt pool affect the mechanical properties of the part, i.e., the formation of porosity and rough surface [49,70,71]. Melt pool size and shape are considered selection factors for other parameters. For instance, Aboulkhair et al. [71] considered melt pool size and shape for Ti64 related to forming pores and surface conditions. They suggested the operator find optimum values of other parameters because the measured porosity result shows an increase at too large or a decrease at too small a beam spot size [33].

Most of the parameters in the laser aspects are interrelated and too complex to figure out as they are very fast interactions with powder morphology timescales. However, energy density or specific energy encapsulates all other characteristics in a single variable. The proper energy density is critical for the melting of materials [26]. The energy density can be described by its spot size, pulse duration and frequency [62]. The fluctuation of energy density causes a repeatability problem in SLM as an inconsistent melt pool has effects, such as excess vaporization, that could occur from the transient higher energy levels [49].

The effectiveness of energy density can be affected by the morphology of powder [57] and scanning-related or printing chamber parameters. For instance, increases and decreases in scanning cause the formation of irregular and metallurgical morphology porosity, respectively [71]. The other scanning-related parameters include the scan speed, scan spacing and scan pattern [49]. There are also printing chamber-related factors, such as atmosphere (inert gas), oxygen level and chamber pressure [72], as well as temperature-related parameters (powder bed temperature, powder bed feeder and temperature uniformity) [49]. Adjusting the process parameters for the powders, chamber environment and laser, one could possibly alter the cooling rate and melt pool size. The important relation is that the quality of the final part is the function of the involving process parameters from all aspects. Ladani et al. [15] asserted the final microstructure of the part is a function of global-level parameters such as build atmosphere, build orientation, scanning strategy and hatch spacing. This infers that the critical combination of those global process parameters can contribute to the quality of the final part.

4.3. Aspects of the Feedstock Material-Related Parameters

Both in traditional manufacturing and AM, feedstock materials are one of the most vital factors in the quality of the finished product. Reduced mechanical qualities or even premature failure of the manufactured product could be caused by the poor quality of raw materials. In principle, all materials that can be melted and re-solidified can be used in the PBF process as feedstock [12]. According to Gibson et al. [19], any metal that can be welded or cast could be used in the SLM process. That is due to the similarities of involving process parameters [62]. However, the generalization of feedstock material selection for SLM based on their characteristics of weldability and castability as a good candidate for SLM does not have a strong scientific background due to the existence of huge differences between welding and casting with the SLM process. Aboulkhair et al. [62] investigated the difference associated with the solidification rate of both the SLM and casting process. The finding shows the solidification rate of the SLM process is much steeper than the casting process. Various mechanical properties and quality indicators, such as microstructures, are affected by the solidification process. According to Wu et al. [73], conventional processing such as welding and casting yields a coarse microstructure compared to SLM. Siddique et al. [74] reported the yield strength of an SLM AlSi12 is four times that of the sand-cast alloy, which they attributed to the fine microstructure. This indicates there could be variables that affect the microstructure of the part, possibly related to the solidification rate. Moreover, the feedstock format used in the case of SLM is powder, while welding uses bulk or wire that affects the laser power absorption behaviors, which is quite similar to the statement under the powder morphology aspect.

Furthermore, the materials repeatedly solidify and melt during re-scanning or re-melting in SLM, which is different from the condition that the same material will go through during welding. Therefore, the laser radiation can affect the alloying elements [61], which will have another impact on the final part, so it is crucial to develop novel alloy systems that are well suited for the SLM in terms of the desirable physical and chemical properties. However, only a few alloys, such as AlSi10Mg, TiAl6V4, CoCr and Inconel 718, are currently the most important and can be successfully printed, claimed by Martin et al. [17] in their study. The majority of more than 5500 alloys in use today cannot be additively manufactured because the melting and solidification dynamics during the printing process led to intolerable microstructures with large columnar grains and periodic cracks [17]. Understanding a material's thermal behavior during SLM is crucial before beginning the experimental qualifying of the material to predict the microstructures and mechanical properties of the parts [62]. However, fusion-based AM of metallic components is a promising technology with plenty of benefits and applications. The challenge from a feedstock perspective is that each application scenario imposes unique requirements on the alloy properties. In this respect, fusion-based AM is still in its infancy, with many available materials not designed for fusion-based AM processes.

The metals used in both AM and conventional manufacturing need parallel consideration of the process parameters involved in the process, i.e., significant alterations to the quality of the part. Table 1 reveals some of the literature that reported the same material processed under different manufacturing process yields with different final part characteristics. It indicates the weakness of the generalization that materials used in welding or casting could be for SLM.

Table 1. Tensile properties of Ti-6Al-4V fabricated by different processes.

Response Parameters	Tensile Properties (Ti-6Al-4V)					
	SLM	**LBW**	**WAAM**	**Casting**	**EBW**	**Units**
Ultimate tensile strength	1267 [68,75]	974 [75]	988 [75]	976 [68]	967 [75]	MPa
Yield strength	1110 [68,75]	942 [75]	909 [75]	847 [68]	922 [75]	MPa
Fracture tensile strain	7.28 [68,75]	10.03 [75]	7.5 [75]	5.1 [68]	9.7 [75]	%
Response parameters	**Hardness and compressive properties (CP-Ti)** [68]					
	LPBF	55% Cold rolled		Casting		Units
Hardness	261	268		210		VH
Compressive strength	1136	900		820		MPa
Maximum strain	51	35		60		%

EBW: Electron beam welding, WAAM: Wire arc additive manufacturing, LBW: laser beam welding, VH: Vickers harness.

5. Powder Bed Fusion Additive Manufacturing Process Parameters and Defects

SLM has a unique ability to efficiently make complex shaped, hollow, thin-walled, or slender parts. The ability to produce a functional part and near-net shape using SLM without post-processing, i.e., machine finishing, would be an ideal use of the SLM process. However, finish machining operations are required to guarantee part assembly and the desired surface quality [76]. For instance, the milling operation, which is a commonly used machining operation for the post-processing of additively produced parts, often encounters challenges due to the anisotropic and crystalline textures. As a result, examining the effects of SLM parameters on the cutting force and anisotropy of the Inconel 718 parts has been the subject of recent research [77]. According to Perez-Ruiz et al. [77], post-machining processes are still necessary to improve dimensional and surface quality, particularly in low-stiffness components. To increase the stiffness of SLM components, they suggested an iterative design process to obtain improved surface roughness and cutting force for thin-walled bent ducts.

The process parameters of the PBF process have great impacts on the final products. Understanding AM defects is essential for failure analysis, defect-based mechanical performance modeling and the structural integrity of load-bearing components [60]. The parameters could be from the powder morphology, laser related, the feedstock material or the printing chamber environments. Currently, the main challenge of PBF is a complete understanding of the relation between the processing parameters and the final quality of components [68]. As mentioned in an earlier section, the factors from different aspects contributed to the final qualities. The output parameters represent the quality of SLM products manifested in the form of geometry, microstructure, surface texture (lay, roughness, waviness), porosity, cracking, distortion, delamination, balling, residual stress [27,43] and others. According to Maurya et al. [27], the defects in AM are common, and that is due to the process parameter; optimization is very important. Therefore, determining which parameters are involved in the formation of specific defects is important.

5.1. Microstructural Defects in PBF-AM Fabricated Parts

The microstructure of the part is vital due to its significant impact on corrosion resistance [78], fracture, toughness [79], ductility [80], thermal conductivity [81] and other prominent properties [13]. Therefore, having consistent microstructure in additively manufactured parts is a big concern. The microstructure is the function of various process parameters, and a variety of microstructures can be produced under different parameters. The fact that each layer is created by partially melting and solidifying the previous layer adds complexity to the SLM process. In addition, SLM experiences a fast cooling rate of 10^4 to 10^6 K/s [46] due to the substantial temperature gradient and even leads to cracks owing to rapid shrinkage [82].

The role of process parameters in forming a variety of microstructures is significant. For instance, laser power, scan speed, layer thickness, hatching distance, scanning strategy and others show different microstructures, including single crystalline-like microstructures, crystallographic lamellar microstructures and polycrystalline microstructures [83]. Crystal textures cause the anisotropy of mechanical properties such as Young's modulus, yield stress and wear resistance [84]. Anisotropic microstructures are commonly developed based on the cooling direction within a melt pool. On the other hand, SLM is highly effective at controlling the crystallographic texture of a wide range of metallic materials, from randomly oriented polycrystalline to single crystalline-like microstructures. The single-crystal-like metal structures are applied to fabricate nearly totally dense tungsten parts with a relative density of 99.1% by tuning the laser process parameters, which is the greatest value yet reported [85]. However, suppression of the occurrence of compositional supercooling is required to obtain a single crystal [84].

Several authors considered the impacts of different process parameters on the condition of the microstructure of metal-based PBF. For instance, Niendorf et al. [86] produced a highly anisotropic microstructure of 316L stainless steel with strong texture by variation of the laser power at the value of 400 and 1000 W, average PSD of 40 μm at 50 μm and 150 μm layer thickness. The use of a 1000 W high-energy laser system reveals a coarse and strongly textured microstructure. The significance of the other parameters (except for laser power at 100 W), such PSD and scanning speed, are not clearly described or emphasized. Martin et al. [17] indicated the contribution of PSD on the formation of microstructures, and the parameters of powder have a significant contribution to altering the form of microstructures. Even the introduction of nanoparticles to SLM is part of the evolution of an isotropic microstructure with fine equiaxed grains [17]. The study on the Inconel 718 samples with layer thicknesses (20, 30, 40 and 50 μm) reveals that the lower the layer thickness, the denser and better the dimensional accuracy [87], which is marginally related to mechanical properties, i.e., microstructure.

Furthermore, several authors considered the energy density or specific energy density as the vital parameter to determine the microstructure of the SLM part. The relationship between specific energy density, microstructure, and corrosion resistance of the SLM specimens of CoCrMo under SLM was investigated [88]. The findings demonstrate that a higher specific energy density promotes columnar grain development and results in coarse grain size, and critical energy density is suggested. Energy density is dependent on other parameters, such as laser power and scanning speed, i.e., increasing scanning speed reduces line energy density [15]. This is due to local energy losses, such as the interaction of the beam with a different morphology of powder and hatching strategy. In addition, a higher beam current causes a loss in alloying elements from the original composition and forms defects in the microstructure. For instance, the research of Ge et al. [89] on Ti-47Al-2Cr-2Nb shows at 10 mA, there was a 15% loss of aluminum, and the microstructure also exhibited a variance with the beam current compared to raw powder composition, as shown in Table 2; the perspective microstructure is illustrated in Figure 3. Therefore, considering the melting point of alloying elements of feedstock materials while applying beam currents is critical.

Table 2. Chemical composition of as-built samples (atomic percentage) [89].

Sample No	Melting Beam Current (mA)	Ti	Al	Cr	Nb
1	4	45.25	50.48	2.17	2.10
2	6	55.63	39.71	2.08	2.58
3	8	58.13	37.43	1.62	2.82
4	10	61.85	33.50	1.47	3.19

In high-tech industries such as energy and aeronautics, some components need to be very lightweight without compromising their strength, which has a direct relation with the microstructure [90]. The microelement size, shape and topology design and optimization are becoming vital challenges, particularly in the manufacturing of skeletons with cavities, which are the concern of additive manufacturing. Calleja-Ochoa et al. [90] proposed the use of "replicative" structures in various sizes and orders of magnitude to manufacture parts with minimum weight while maintaining the needed mechanical properties.

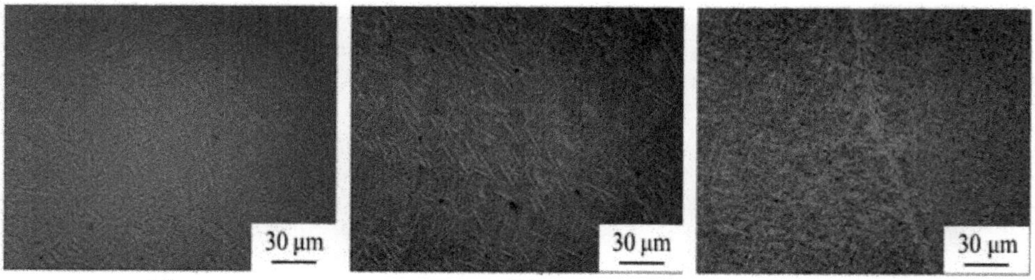

(a) (b) (c)

Figure 3. Optical microscopy in top region (**a**) Sample-2, (**b**) Sample-3 and (**c**) Sample-4 (Reused by permission form [89], License number 5462390275210).

In general, the literature shows that the microstructure is the actual concern of SLM, and scholars identified process parameters that need to be monitored separately. The property of the material under consideration is also essential since the dynamics of melting and solidification during the printing process result in intolerable microstructure and cracks, as already mentioned by Martin et al. [17]. However, there is small coverage with collective impacts that are related to laser, powder morphology, feedstock material and chamber condition.

5.2. Surface Roughness of PBF of Produced Parts

Surface texture is the geometrical irregularities or roughness that exist on the surface [57] and is influenced by different factors of SLM [27,43]. Surface quality is one of the unsolved problems in SLM [91]. The surface finish condition plays a key role in the mechanical, tribological and functional performance of SLM components [43]. For example, an increase in surface roughness tends to cause faster crack initiation, with a resulting decrease in fatigue performance [92].

The surface roughness of SLM parts is nearly four to five times of the machined surface. Therefore, addressing those responsible process parameters is the principal endeavor. From the powder morphologies, PSD, shape, smoothness, apparent density and flowability are some of the factors considered by various authors that have shown significant contributions to surface roughness [48,56,57,93]. The large particles inside layers could cause a rough surface because the volume of particles tends to form voids when they are removed in the finishing process [94]. The staircase is another source of a rough surface due to the layered nature of part construction, creating a stair-stepping effect that is heavily dependent on part orientation [95]. In addition to surface roughness, the staircase causes dimensional inaccuracy and volumetric error, as shown in Figure 4, where Si is the contour area of *i*th layer.

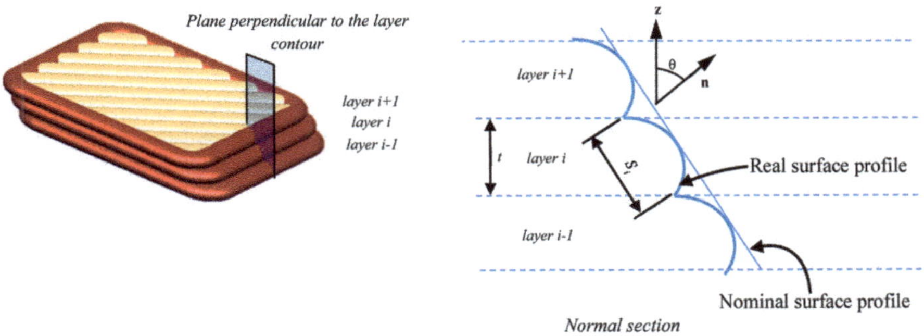

Figure 4. SLM process and staircase error (Reused from [96] with permission in accordance with Creative Common CC BY license, open access).

The other surface roughness challenge of SLM comes from the standard triangulation language (STL) file [97], as it is responsible for creating slice thickness that leads to a staircase effect on the part. The variation of surface roughness with the .STL file is given in Table 3. In the other case, the combined effect of the scanning speed and layer thickness study shows as both parameters increase, the surface roughness increases, which is given in Table 4. However, the study missed that layer thickness could be affected by the flowability of the powder. Poor powder flowability blocks the spreading of particles and affects the continuity of layer-led rough surfaces [43]. This implies the powder morphology, which determines the flowability status, has indirect impacts on the surface condition of the component.

Table 3. Surface roughness and deviation [97].

Sample	No. of Triangles	Ra [μm]	Rz [μm]	Deviation .STL File
1	40,242	17.2	90.0	0.0286
2	1,994,398	19.5	101.2	0.0231
3	69,642	12.4	61.2	0.0258
4	40,242	17.9	87.4	0.0292
5	69,698	21.2	102.2	0.0331
6	1,994,398	18.4	90.9	0.0311
7	6114	17.8	82.2	0.0289
8	69,698	20.2	106.7	0.0314
9	11,304	20.4	95.1	0.0316
10	4880	20.6	109.3	0.0325
11	40,242	18.9	80.4	0.0318
12	1,994,398	15.2	79.3	0.0271
13	1,994,398	16.9	77.4	0.0282
14	10,502	14.6	88.3	0.0268
15	10,502	17.5	85.6	0.0285
16	4516	24.2	124.1	0.0347

Table 4. Surface roughness (Ra) at different layer thicknesses and scanning speeds [56].

Roughness, μm	Surface	30 μm	50 μm	70 μm
Scanning speed = 70 mm/s	Top	25.67	29.8	35.9
	Side	15.67	18.6	18.9
Scanning speed = 90 mm/s	Top	26	34	41.6
	Side	16.8	19.8	20

The interdependency of the surface quality and the laser process parameters, although it is mentioned in the literature, is still not yet fully analyzed or quantified. According to Galy et al. [98], the usually adjusted parameters for surface improvement are those from energy input into the part contour. The surface variation often results from the adjustment of linear energy density consisting of laser power and scan speed, which is given by Equation (1). The Ψ is specific laser energy (J/mm^3), P is the power of the laser (W), v is the scan speed of the laser (mm/ s), h is the inter-bead distance (mm) and d is the thickness of the layer of powder deposited (mm) [98].

$$\psi = \frac{P}{v.h.d} \tag{1}$$

However, as many researchers debate on the equation, the other parameters, such as laser spot diameter, the direction of gas flow and hatching strategies, are not deemed. This is due to the intensity of laser energy being varied by focal position via laser beam diameter on the build platform [92]. So, the interdependency of those parameters becomes a vital factor for the instability of melt pool size and shape and leads to balling, pores and variation on the up skin of the component. Chowdhury et al. [68] found that the tendency of ball formation can be reduced by increasing energy density. However, the excessive input energy causes the spreading and melting of the previous layer [46], which leads to another issue on the microstructure and forms complexity to other properties. In general, the common source of surface roughness can come from the powder, material and working chamber. However, the limited literature coverage is given for the factors from the chamber that includes the inert gas and limited oxygen. According to Li et al. [91], although shielding gas may prevent oxidization, it is very hard to completely eliminate the oxidation when the experiments were carried out in the air, especially for some active metals element.

5.3. Porosity of PBF-Produced Parts

One of the most common defects in SLM parts in general, irrespective of the material, is porosity, so more studies focus on the mechanism, causes and minimization of porosity [62,71,99–101]. According to Coro et al. [102], AM technology is an excellent choice for future designs. However, if the printing parameters are not fine-tuned, pores can emerge during the process. Due to the steep cooling rate process in the SLM, the cracking of parts is expected [101]. The presence of porosity potentially promotes cracking due to the notch effect [8,27,103–106]. Pores can develop over time due to entrained gas, chemical reactions, and supersaturated dissolved gases that create air holes in the melt pool [107]. The input process parameters responsible for the formation of porosity are suggested in the previous work of researchers. For instance, Peng et al. [101] developed the energy demand model to manufacture the SLM steel parts using critical parameters such as laser power, scan speed, layer thickness and hatch spacing.

The porosity in SLM can manifest on the surface or sub-surface. The sub-surface porosities are mostly located in the transitional zone between the counter path and the hatching path [8]. The porosity in the SLM is categorized into spherical (gas or hydrogen porosity) and irregular (keyhole or lack-of-fusion) morphology [27,72]. The size and shape of porosity indicate the phenomenon that has caused it [98]. The presence of moisture on the powder surface can easily capture hydrogen and contributes to the spherical morphology of pore development [62], which is known as metallurgical pores. In the other case, the presence of non-molten powder causes a lack of fusion, or keyhole pores are the categories under irregular morphology. According to Tang et al. [100], oxides formed during the melting and solidification of the sample are some of the conditions responsible for the formation of irregular morphology. This implies the parameters in the melting and solidification process contribute to the formation of irregular morphology porosities. The indirect description for porosity is the relative density of the part. Components that have lower relative density or higher pores are exposed to premature failure [107].

From the wing of laser parameters, most studies on reducing the rate of porosity within a material produced by SLM focus on the process and especially on the optimization of the energy density factor. As mentioned earlier, energy density is the function of laser scan speed and laser power. Kimura et al. [107] investigated how energy density affects the development of pores. So, the finding suggests that the part's relative density dropped at both greater and lower energy densities. Incomplete melting of the powder causes porosity at lower energy densities, and at higher densities, many spherical porosities appear as morphological features of gas pores, either trapping argon from the combustion chamber due to the active flow of the molten metal or because the dissolved hydrogen aired during fusion. The relative density of A356.0 SLM specimens created with various levels of laser power at various scan speeds at a constant scan spacing of 0.1 mm is depicted in Figure 5a. The density of the specimens is reduced with increasing or decreasing scan speeds. For the specimens created at a specific scan speed and laser power setting, a peak density was discovered. The relative density of the specimens plotted as a function of scan spacing is shown in Figure 5b. In addition, the study conducted on Ti-6Al-4V fabricated by SLM indicates that as the interaction between high laser beam and powder particles at fixed layer thickness (20 μm), the samples contain low porosity when the speed is below 2700 mm/s [98]. Further increasing the speed led to an increase in porosity but not a significant one. Gong et al. [108] suggested the operator find optimum values to balance the parameters due to the result of his measured porosity showing increments of either too large or too small a beam spot size.

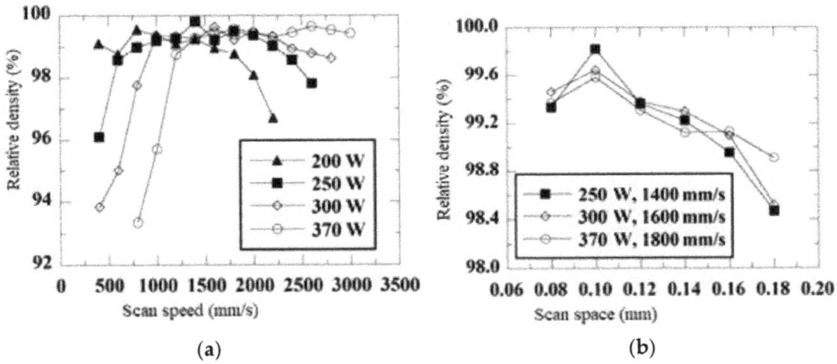

Figure 5. Relative density of SLM as (**a**). A function of scan speed and laser power, (**b**). Scan spacing. (Reused by permission form [107], License number 5464270902364).

In SLM, melting and solidification occur locally; the cooling rate is the function parameters such as beam power, beam speed, the pre-heat condition, build geometry and hatch spacing. Koutiri et al. [8] investigated the possible link between volume energy density and the porosity rate of SLM by incorporating the hatching distance with laser power and scanning speed. Two geometries of pores are observed as circular, which is close to the up skin and the irregular one, i.e., the keyhole with deeper beads formed around the overlap between the two strategies (contour and hatching). The dependence of porosity on those parameters is shown in Figure 6.

In the other case, higher relative density in powders improves the process by reducing internal stresses, part distortion and final part porosity [109]. This is due to the thermal conductivity increment with a relative density of the particle. The effects of powder particle size, shape and distribution on the porosity of the SLM part is another aspect. However, several researchers considered parameters related to energy density, such as laser power, scan speed, scan spacing and layer thickness [57,110,111]. The material absorptivity of laser energy and powder particle size needs to be considered to analyze the porosity formation rate.

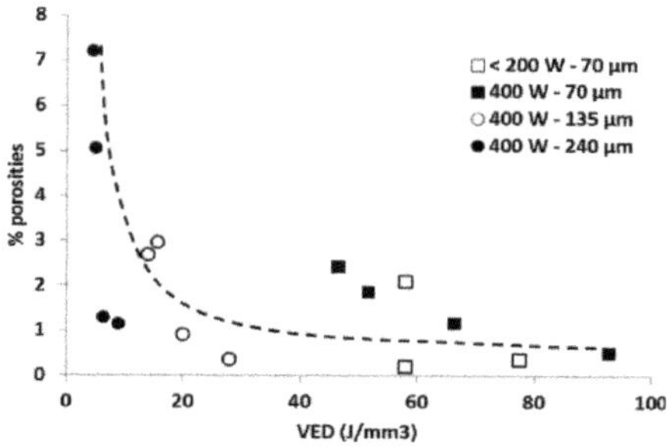

(a)

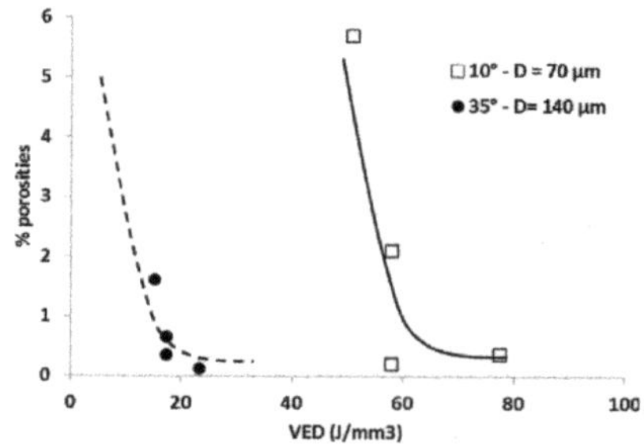

(b)

Figure 6. Dependence of porosity content versus process parameters in In625 for (**a**). P < 200 W and P = 400 W (10°), (**b**). Two distinct building angles (10° and 35°) (Adapted from [8] with permission, License number 5462380625001).

5.4. Residual Stress in PBF-Produced Parts

High residual stresses are one of the fundamental phenomena that hinder SLM's widespread adoption in the industry, even though it is now a well-established metal manufacturing method. The stress is created by the large thermal gradients created by the laser that passes over an area with the powder rapidly melting, then back to solidification [18,49,62,112]. As the name suggests, these stresses remain in a component once the material has come to equilibrium with the environment [112,113]. Some of the effects of residual stresses are warping and thermal stress-related cracking, which cannot be corrected by post-processing [113]. Shrinkage is another form of shape distortion, which causes a lack of dimensional accuracy [112].

The exaggerated residual stress in SLM [62,112] inhibits the application of parts due to the aforementioned flaws. Therefore, it is crucial to identify input process parameter combinations that produce the smallest residual stress and associated defects magnitudes. The mechanism of residual stresses can be due to the temperature gradient that occurs at the laser spot and cooling rate. The mechanisms of residual stresses in SLM are well explained in [112]. The major defects due to residual stresses are dimensional deviation [114], delamination, stress-induced cracking and accelerated crack growth, which leads to early fatigue failure [115]. Warping, distortions and cracking are irreversible by post-processing techniques such as stress relief heat treatment [116,117]. Therefore, due to the catastrophic failure of components, understanding the process parameters responsible for the cause of residual stress is crucial for mitigation in advance. The negative impacts of residual stresses can be greatly diminished by manipulating one or more of the process parameters. As mentioned by Mugwagwa et al. [115], some of the factors that influence residual stresses include scanning patterns, scan vector lengths, scan vector angles, rotation angles between layers, part geometry, material type, support types and pre-heating conditions. The residual stress formation and process parameter mapping of SLM needs deep insight and still require more consideration.

Several authors identified influential process parameters responsible for the induced residual stress. For instance, Yadroitsev et al. [117] considered laser power, layer thickness and scanning speed to analyze the residual stress of SLM, while several researchers considered laser power and scanning speed [118–120]. The effect of process gas, scan speed and sample thickness on the buildup residual stresses and porosity in Ti-6Al-4V produced by SLM was studied [121]. In the other case, the common defect in SLM porosity has a positive effect on the residual stress [115,122,123]. That is the trick that needs to be emphasized in the research to clarify at what level of porosity is tolerated to handle the residual stress without exposing the components because of the porosity. The scatter plot in Figure 7 shows that pores indeed have the effect of relaxing residual stresses, and higher porosity leads to less distortion. The interdependence of porosity and residual stress will give consideration to both influential process parameters. In several studies, the constraints from powder particles and individual influential process parameters for distortion, residual stress and porosity are not clearly considered, and the focus of the studies was to reveal the relationship among the responses.

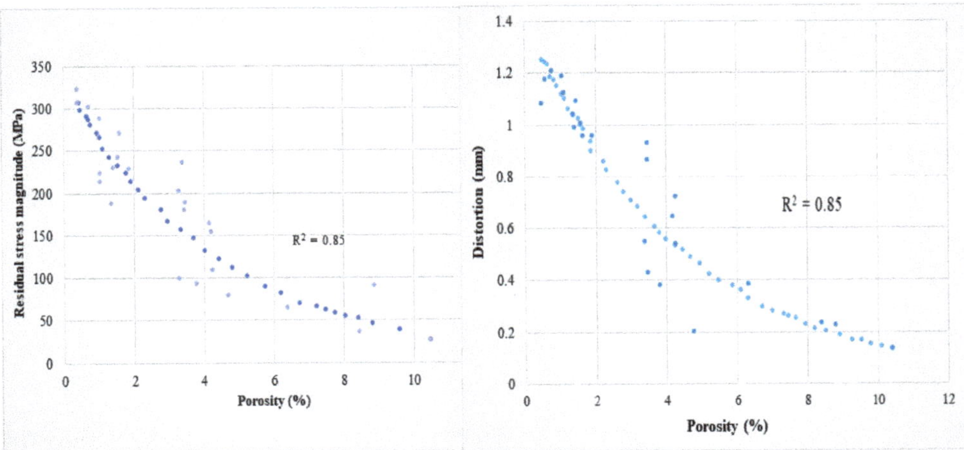

(a) (b)

Figure 7. SLM interdependencies: (**a**) Porosity residual stresses, (**b**) Porosity distortions (Adapted from [115] with permission in accordance with Creative Common CC BY license, open access).

The influence of energy density on the residual stress is studied. When parameters are varied simultaneously, there is no observable influence of energy density on residual stresses, as shown in Figure 8. Thus, according to Mugwagwa et al. [115], energy density cannot be used to explain or account for the differences in the observed process outcome, which is similar to the observation of Prashanth [58]. However, the energy density is the function of laser power and scanning speed; considering those parameters, one factor may affect the result.

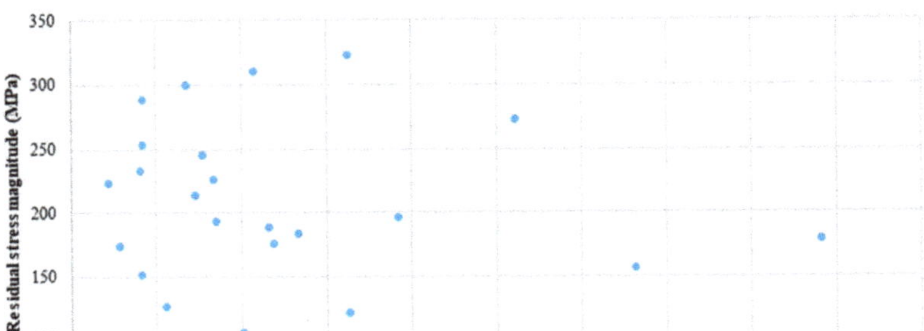

Figure 8. Scatter plots of the effect of energy density on residual stress magnitude (Adapted from [115] with permission in accordance with Creative Common CC BY license, open access).

5.5. Repeatability and Reproducibility Issues of PBF Process

According to the review article published in 2014 by Tapia et al. [6], a consensus among experts and stakeholders in the aerospace, healthcare and automotive fields indicate that metallic AM parts are still not sufficient to meet their strict requirements. The high-tech industries demand overall equipment effectiveness (OEE), which must be better than 70%, with scrap rates of less than 1000 ppm as a crucial criterion. The current world-class vehicle manufacturer's OEE is increased to 85% [49]. Weller et al. [124] further demonstrated the consensus in a comprehensive economic review on the implications and limitations of AM in the industry, and the greatest barrier is identified as the limitation of repeatability and reproducibility of parts, which affects the OEE calculations. According to ASTM E117, repeatability is precision under the conditions where independent test results are obtained with the same method on identical test items in the same laboratory by the same operator using the same equipment within short intervals of time [49]. Reproducibility is similar, but the test can be carried out in different laboratories, with different operators and using different equipment. The high volume of research has focused on the capability of AM, but there is a lack of research coverage focused on its repeatability and reproducibility.

The product quality and reliability characteristics, i.e., dimensional accuracy, precision, repeatability and reproducibility, are mostly affected by inherent and systematic manufacturing process variations [18]. One of the common errors that occur in PBF is dimensional accuracy. Researchers focused on the source of variation in the geometry and mechanical properties of parts. For instance, Franchitti et al. [125] considered orientation angle, build location and build height effects on the dimensional accuracy of Ti-6Al-4V rectangular parts fabricated by the EBM process. Thus, sample orientation and build location were identified as the influential source of dimensional variations. The mechanical precision of the manufacturing setup, such as layer thickness, concentrated laser spot size and scanner's position precision, is among the factors affecting dimensional accuracy [18]. According to Abedi et al. [126] The process variables such as temperature, strain and strain rate have a direct impact on the suggested constitutive equations' precision and repeatability. Calignano et al. [97] investigated the dimensional accuracy of L-PBF using AlSi10Mg alloy and stated that the accuracy of parts produced is affected by the .STL file, build direction and process parameters. The choice of parameters for the .STL file affects not only the accuracy but also the surface roughness of the part's build direction [96] and other process parameters [18]. Here, dimensional repeatability is considered, and the mechanical property of the part is not covered, and that is the unseen part of the literature. However, researchers investigated the repeatability and reproducibility of PBF through the statistical analysis of standard deviation, where each data set represents different operating parameters. This conflicts with the definition given by ASTM E117.

The issue of repeatability and reproducibility needs the various aspects, powder particles, laser parameters, materials and the printing chamber environments. According to Dawes et al. [55], the more different the PSD in cobalt chrome, the better the ultimate tensile strength obtained as the size of the powder increases; however, the repeatability of these parts is reduced. This implies the PSD has contributed to altering the repeatability of the part. The flowability of the powder has direct effects on the supplied layer thickness of parts and yields different yield strengths and UTS [49].

In the other case, the consequence of other defects like residual stress can have effects on the repeatability of the part. The volume shrinkage and thermal gradients together lead to warpage and distortion, which causes dimensional inaccuracies and loss of quality in the final part [96]. The effects of stair-stepping due to the layered nature of part construction which is heavily dependent on part orientation [94], in addition to surface roughness, can cause dimensional inaccuracies.

6. Conclusions

In this article, a review of the recent literature on research advances and applications of powder-bed fusion-based additive manufacturing technology is conducted. The study focused on SLM-based AM technology, and based on the review, the following conclusions are observed.

- In general, the current status of the SLM production process indicates that it is still expensive and slow compared to the conventional manufacturing process. The components are typically semi-finished and need post-processing.
- PBF-based metal AM received attention from R&D due to its rewards, such as a high degree of product customization and minimum buy-to-fly ratio. Moreover, the SLM has the capacity to manufacture materials that would be difficult by other manufacturing processes.
- The comparison of PBF and its sub-categories with other related processes, such as DED, are among the concerns of the industries. The status of PBF indicates that SLM is an emerging candidate for mission-oriented applications like aerospace, defense and healthcare, while EB-PBF is a newer and less explored technology with limited applications.
- SLM suffers from the drawbacks such as reliability and quality in terms of dimensional accuracy, strength and surface roughness.
- The outcome of SLM largely depends on process parameters and the interplay of physical phenomena. As the status and demand of SLM are changing from a rapid prototype to mass production and then to customized production, the underdeveloped issues related to processing parameters are a great future concern of the research for the competence of the technology.
- Mapping process parameters with the outcome is the safest alternative to finding the root cause of variations and monitoring in advance. However, the involved process parameters in the phenomena are complex and change in a timescale. Controlling certain quality aspects of a part could highly affect the unseen aspects. This paper grouped the source of process parameters to powder morphology-related laser parameters, feedstock material properties and printing chamber environments. The finding is a vital input for the bottom-up approach. In addition, considering the relationships of the involved process parameters will lead to fruitful results on the quality of the components.

Author Contributions: Conceptualization, N.D.D. and H.G.L.; methodology, N.D.D.; software, N.D.D.; validation, H.G.L.; formal analysis, N.D.D.; investigation, N.D.D.; resources, H.G.L.; data curation, N.D.D.; writing—original draft preparation, N.D.D.; writing—review and editing, H.G.L.; visualization, N.D.D. and H.G.L.; supervision, H.G.L.; project administration, H.G.L.; funding acquisition, H.G.L. All authors have read and agreed to the published version of the manuscript.

Funding: This research was supported by the INDMET project, grant number 62862, funded by the NORHED II program.

Institutional Review Board Statement: Not applicable.

Data Availability Statement: Not applicable.

Conflicts of Interest: The authors declare no conflict of interest.

References

1. Colosimo, B.M.; Huang, Q.; Dasgupta, T.; Tsung, F. Opportunities and challenges of quality engineering for additive manufacturing. *J. Qual. Technol.* **2018**, *50*, 233–252. [CrossRef]
2. Mehrpouya, M.; Tuma, D.; Vaneker, T.; Afrasiabi, M.; Bambach, M.; Gibson, I. Multi-material powder bed fusion techniques. *Rapid Prototyp. J.* **2022**, *28*, 1–19. [CrossRef]
3. Dilberoglu, U.M.; Gharehpapagh, B.; Yaman, U.; Dolen, M. The Role of Additive Manufacturing in the Era of Industry 4.0. *Procedia Manuf.* **2017**, *11*, 545–554. [CrossRef]
4. Dovgyy, B. Assessing the Printability of Alloys in Fusion-Based Additive Manufacturing: Towards Criteria for Alloy Selection. Doctoral Dissertation, Imperial College: London, UK, 2022.
5. Lewis, G. Aspects of the Powder in Metal Additive Manufacturing: Review. *World J. Text. Eng. Technol.* **2022**, *10*, 363–409. [CrossRef]

6. Tapia, G.; Elwany, A. A review on process monitoring and control in metal-based additive manufacturing. *J. Eng. Ind.* **2014**, *136*, 60801. [CrossRef]
7. Babuska, T.F.; Krick, B.A.; Susan, D.F.; Kustas, A.B. Comparison of powder bed fusion and directed energy deposition for tailoring mechanical properties of traditionally brittle alloys. *Manuf. Lett.* **2021**, *28*, 30–34. [CrossRef]
8. Koutiri, I.; Pessard, E.; Peyre, P.; Amlou, O.; De, T.T. Influence of SLM process parameters on the surface finish, porosity rate and fatigue behavior of as-built Inconel 625 parts. *J. Mater. Process. Technol.* **2018**, *255*, 536–546. [CrossRef]
9. Chen, Z.; Han, C.; Gao, M.; Kandukuri, S.Y.; Zhou, K. A review on qualification and certification for metal additive manufacturing. *Virtual Phys. Prototyp.* **2022**, *17*, 382–405. [CrossRef]
10. Wohlers, T.; Gornet, T. History of additive manufacturing. *Wohlers Rep.* **2014**, *24*, 118.
11. Lee, J.Y.; An, J.; Chua, C.K. Fundamentals, and applications of 3D printing for novel materials. *Appl. Mater. Today* **2017**, *7*, 120–133. [CrossRef]
12. Gibson, I.; Rosen, D.; Stucker, B.; Khorasani, M. Design for additive manufacturing. In *Additive Manufacturing Technologies*; Springer: Cham, Switzerland, 2021; pp. 555–607.
13. Singh, R.; Gupta, A.; Tripathi, O.; Srivastava, S.; Singh, B.; Awasthi, A.; Rajput, S.K.; Sonia, P.; Singhal, P.; Saxena, K.K. Powder bed fusion process in additive manufacturing: An overview. *Mater. Today Proc.* **2020**, *26*, 3058–3070. [CrossRef]
14. Garcia-Colomo, A.; Wood, D.; Martina, F.; Williams, S.W. A comparison framework to support the selection of the best additive manufacturing process for specific aerospace applications. *Int. J. Rapid Manuf.* **2020**, *9*, 194–211. [CrossRef]
15. Ladani, L.; Sadeghilaridjani, M. Review of powder bed fusion additive manufacturing for metals. *Metals* **2021**, *11*, 1391. [CrossRef]
16. Murugan, P.D.; Vijayananth, S.; Natarajan, M.P.; Jayabalakrishnan, D.; Arul, K.; Jayaseelan, V.; Elanchezhian, J. A current state of metal additive manufacturing methods: A review. *Mater. Today Proc.* **2021**, *59*, 1277–1283. [CrossRef]
17. Martin, J.M.; Yahata, B.D.; Hundley, J.M.; Mayer, J.A.; Schaedler, T.A.; Pollock, T.M. 3D printing of high-strength aluminum alloys. *Nature* **2017**, *549*, 365–369. [CrossRef]
18. Ahmed, A.; Majeed, A.; Atta, Z.; Jia, G. Dimensional quality, and distortion analysis of thin-walled alloy parts of AlSi10Mg manufactured by selective laser melting. *J. Manuf. Mater. Process.* **2019**, *3*, 51. [CrossRef]
19. Gibson, I.; Rosen, D.; Stucker, B.; Khorasani, M. *Additive Manufacturing Technologies*, 3rd ed.; Springer: Cham, Switzerland, 2021; Volume 17.
20. Džugan, J.; Nový, Z. *Powder Application in Additive Manufacturing of Metallic Parts. Powder Metallurgy: Fundamentals and Case Studies*; InTechOpen: Rijeka, Croatia, 2017; p. 183.
21. Ahmadi, M.; Tabary, B.; Rahmatabadi, D.; Ebrahimi, S.; Abrinia, K.; Hashemi, R. Review of selective laser melting of magnesium alloys: Advantages, microstructure and mechanical characterizations, defects, challenges, and applications. *J. Mater. Res. Technol.* **2022**, *19*, 1537–1562. [CrossRef]
22. Yang, L.; Hsu, K.; Baughman, B.; Godfrey, D.; Medina, F.; Menon, M.; Wiener, S. *Additive Manufacturing of Metals: The Technology, Materials, Design, and Production*; Springer: Cham, Switzerland, 2017; pp. 45–61.
23. Vafadar, A.; Guzzomi, F.; Rassau, A.; Hayward, K. Advances in metal additive manufacturing: A review of common processes, industrial applications, and current challenges. *Appl. Sci.* **2021**, *11*, 1213. [CrossRef]
24. Korpela, M.; Riikonen, N.; Piili, H.; Salminen, A.; Nyrhilä, O. Additive manufacturing—Past, present, and the future. In *Technical, Economic, and Societal Effects of Manufacturing 4.0*; Palgrave Macmillan: Cham, Switzerland, 2020; pp. 17–41.
25. King, W.E.; Anderson, A.T.; Ferencz, R.M.; Hodge, N.E.; Kamath, C.; Khairallah, S.A.; Rubenchik, A.M. Laser powder bed fusion additive manufacturing of metals; physics, computational, and materials challenges. *Appl. Phys. Rev.* **2015**, *2*, 041304. [CrossRef]
26. Wang, Y.; Chen, X.; Jayalakshmi, S.; Singh, R.A.; Sergey, K.; Gupta, M. Process parameters, product quality monitoring, and control of powder bed fusion. In *Transactions on Intelligent Welding Manufacturing*; Springer: Singapore, 2020; pp. 89–108.
27. Maurya, A.K.; Kumar, A. Defect and Distortion in Powder Bed Fusion of Metal Additive Manufacturing Parts. *ASEAN J. Sci. Technol. Dev.* **2022**, *39*, 85–103. [CrossRef]
28. Milewski, J.O. *Additive Manufacturing of Metals*; Springer International Publishing AG: Cham, Switzerland, 2017; Volume 258, pp. 134–1577.
29. Chua, Z.Y.; Ahn, I.H.; Moon, S.K. Process monitoring and inspection systems in metal additive manufacturing: Status and applications. *Int. J. Precis. Eng. Manuf.-Green Tech.* **2017**, *4*, 235–245. [CrossRef]
30. Childerhouse, T.; Jackson, M. Near net shape manufacture of titanium alloy components from powder and wire: A review of state-of-the-art process routes. *Metals* **2019**, *9*, 689. [CrossRef]
31. Dutta, B.; Froes, F.S. The additive manufacturing of titanium alloys. *Met. Powder Rep.* **2017**, *72*, 96–106.
32. Pal, R.; Basak, A. Linking Powder Properties, Printing Parameters, Post-Processing Methods, and Fatigue Properties in Additive Manufacturing of AlSi10Mg. *Alloys* **2022**, *1*, 149–179. [CrossRef]
33. Francis, Z.R. The Effects of Laser and Electron Beam Spot Size in Additive Manufacturing Processes. Doctoral Dissertation, Carnegie Mellon University, Pittsburgh, PA, USA, 2017.
34. Baumers, M.; Dickens, P.; Tuck, C.; Hague, R. The cost of additive manufacturing: Machine productivity, economies of scale and technology-push. *Technol. Forecast. Soc. Chang.* **2016**, *102*, 193–201. [CrossRef]
35. Sames, W. Additive Manufacturing of Inconel 718 Using Electron Beam Melting: Processing, Post-Processing, & Mechanical Properties. Doctoral Dissertation, Texas A & M University, College Station, TX, USA, 2015.

36. Ding, D.; Pan, Z.; Cuiuri, D.; Li, H. Wire-feed additive manufacturing of metal components: Technologies, developments, and future interests. *Int. J. Adv. Manuf. Technol.* **2015**, *81*, 465–481.
37. Vayre, B.; Vignat, F.; Villeneuve, F. Metallic additive manufacturing: State-of-the-art review and prospects. *Mech. Ind.* **2012**, *13*, 89–96. [CrossRef]
38. Boley, C.D.; Khairallah, S.A.; Rubenchik, A.M. Calculation of laser absorption by metal powders in additive manufacturing. In *Additive Manufacturing Handbook*; CRC Press: Boca Raton, FL, USA, 2017; pp. 507–509.
39. Gu, D. *Laser Additive Manufacturing (AM): Classification, Processing Philosophy, and Metallurgical Mechanisms*; Springer: Berlin/Heidelberg, Germany, 2015; pp. 15–71.
40. Popov, V.V.; Grilli, M.L.; Koptyug, A.; Jaworska, L.; Katz-Demyanetz, A.; Klobčar, D.; Balos, S.; Postolnyi, B.O.; Goel, S. Powder bed fusion additive manufacturing using critical raw materials: A review. *Materials* **2021**, *14*, 909. [CrossRef]
41. Oliveira, J.P.; Lalonde, A.D.; Ma, J. Processing parameters in laser powder bed fusion metal additive manufacturing. *Mater. Des.* **2020**, *193*, 1–12. [CrossRef]
42. Sow, M.C.; De, T.T.; Castelnau, O.; Hamouche, Z.; Coste, F.; Fabbro, R.; Peyre, P. Influence of beam diameter on Laser Powder Bed Fusion (L-PBF) process. *Addit. Manuf.* **2020**, *36*, 101532.
43. Narasimharaju, S.R.; Zeng, W.; See, T.L.; Zhu, Z.; Scott, P.; Jiang, X.; Lou, S. A comprehensive review on laser powder bed fusion of steels: Processing, microstructure, defects and control methods, mechanical properties, current challenges, and future trends. *J. Manuf. Process.* **2022**, *75*, 375–414. [CrossRef]
44. Montazeri, M.; Rao, P. Sensor-based build condition monitoring in laser powder bed fusion additive manufacturing process using a spectral graph theoretic approach. *J. Manuf. Sci. Eng.* **2018**, *140*, 091002-1. [CrossRef]
45. Zhang, B.; Li, Y.; Bai, Q. Defect Formation Mechanisms in Selective Laser Melting. A Review. *Chin. J. Mech. Eng.* **2017**, *30*, 515–527. [CrossRef]
46. Javidrad, H.R.; Salemi, S. Effect of the volume energy density and heat treatment on the defect, microstructure, and hardness of L-PBF Inconel 625. *Metall. Mater. Trans A* **2020**, *51*, 5880–5891. [CrossRef]
47. Klingbeil, N.W.; Beuth, J.L.; Chin, R.K.; Amon, C.H. Residual stress-induced warping in direct metal solid freeform fabrication. *Int. J. Mech. Sci.* **2002**, *44*, 57–77. [CrossRef]
48. Kladovasilakis, N.; Charalampous, P.; Kostavelis, I.; Tzetzis, D.; Tzovaras, D. Impact of metal additive manufacturing parameters on the powder bed fusion and direct energy deposition processes: A comprehensive review. *Prog. Addit. Manuf.* **2021**, *6*, 349–365. [CrossRef]
49. Dowling, L.; Kennedy, J.; O'Shaughnessy, S.; Trimble, D. A review of critical repeatability and reproducibility issues in powder bed fusion. *Mater. Des.* **2020**, *186*, 108346. [CrossRef]
50. Scime, L.; Beuth, J. Anomaly detection and classification in a laser powder bed additive manufacturing process using a trained computer vision algorithm. *Addit. Manuf.* **2018**, *19*, 114–126. [CrossRef]
51. Kulkarni, P.A.; Berry, R.J.; Bradley, M.S. Review of the flowability measuring techniques for powder metallurgy industry. Proceedings of the Institution of Mechanical Engineers. *J. Process Mech. Eng. Part E* **2010**, *224*, 159–168. [CrossRef]
52. Karapatis, N.P.; Egger, G.; Gygax, P.E.; Glardon, R. Optimization of powder layer density in selective laser sintering. In *1999 International Solid Freeform Fabrication Symposium*; The University of Texas at Austin: Austin, TX, USA, 1999. [CrossRef]
53. Slotwinski, J.A.; Garboczi, E.J.; Stutzman, P.E.; Ferraris, C.F.; Watson, S.S.; Peltz, M.A. Characterization of metal powders used for additive manufacturing. *J. Res. Natl. Inst. Stand. Technol.* **2014**, *119*, 460. [CrossRef]
54. Yuan, B.; Guss, G.M.; Wilson, A.C.; Hau-Riege, S.P.; DePond, P.J.; McMains, S.; Matthews, M.J.; Giera, B. Machine-learning-based monitoring of laser powder bed fusion. *Adv. Mater. Technol.* **2018**, *3*, 1800136. [CrossRef]
55. Dawes, J.; Bowerman, R.; Trepleton, R. Introduction to the Additive Manufacturing Powder Metallurgy Supply Chain Exploring the production and supply of metal powders for AM processes. *Johns. Matthey Technol. Rev.* **2015**, *3*, 243–256. [CrossRef]
56. Spierings, A.B.; Herres, N.; Levy, G. Influence of the particle size distribution on surface quality and mechanical properties in AM steel parts. *Rapid Prototyp. J.* **2011**, *17*, 195–202. [CrossRef]
57. Cherry, J.A.; Davies, H.M.; Mehmood, S.; Lavery, N.P.; Brown, S.G.; Sienz, J. Investigation into the effect of process parameters on microstructural and physical properties of 316L stainless steel parts by selective laser melting. *Int. J. Adv. Manuf. Technol.* **2015**, *76*, 869–879. [CrossRef]
58. Attar, H.; Prashanth, K.G.; Zhang, L.C.; Calin, M.; Okulov, I.V.; Scudino, S.; Yang, C.; Eckert, J. Effect of powder particle shape on the properties of in situ Ti–TiB composite materials produced by selective laser melting. *J. Mater. Sci. Technol.* **2015**, *31*, 1001–1005. [CrossRef]
59. Polozov, I.; Sufiiarov, V.; Kantyukov, A.; Razumov, N.; Goncharov, I.; Makhmutov, T.; Silin, A.; Kim, A.; Starikov, K.; Shamshurin, A.; et al. Microstructure, densification, and mechanical properties of titanium intermetallic alloy manufactured by laser powder bed fusion additive manufacturing with high-temperature preheating using gas atomized and mechanically alloyed plasma spheroidized powders. *Addit. Manuf.* **2020**, *34*, 101374. [CrossRef]
60. Sanaei, D. Characteristics and analysis of their variability in metal L-PBF additive manufacturing. *Mater. Des.* **2019**, *182*, 108091. [CrossRef]
61. Ardila, L.C.; Garciandia, F.; González-Díaz, J.B.; Álvarez, P.; Echeverria, A.; Petite, M.M.; Deffley, R.; Ochoa, J. Effect of IN718 recycled powder reuse on properties of parts manufactured by means of selective laser melting. *Phys. Procedia* **2014**, *56*, 99–107. [CrossRef]

62. Aboulkhair, N.T.; Simonelli, M.; Parry, L.; Ashcroft, I.; Tuck, C.; Hague, R. 3D printing of Aluminium alloys: Additive Manufacturing of Aluminum alloys using selective laser melting. *Prog. Mater. Sci.* **2019**, *106*, 100578. [CrossRef]
63. Simonelli, M.; Tuck, C.; Aboulkhair, N.T.; Maskery, I.; Ashcroft, I.; Wildman, R.D.; Hague, R. A study on the laser spatter and the oxidation reactions during selective laser melting of 316L stainless steel, Al-Si10-Mg, and Ti-6Al-4V. *Metall. Mater. Trans. A* **2015**, *46*, 3842–3851. [CrossRef]
64. Maamoun, A.H.; Elbestawi, M.; Dosbaeva, G.K.; Veldhuis, S.C. Thermal post-processing of AlSi10Mg parts produced by Selective Laser Melting using recycled powder. *Addit. Manuf.* **2018**, *21*, 234–247. [CrossRef]
65. Kamath, C. Data mining and statistical inference in selective laser melting. *Int. J. Adv. Manuf. Technol.* **2016**, *86*, 1659–1677. [CrossRef]
66. Berumen, S.; Bechmann, F.; Lindner, S.; Kruth, J.P.; Craeghs, T. Quality control of laser-and powder bed-based Additive Manufacturing (AM) technologies. *Phys. Procedia* **2010**, *5*, 617–622. [CrossRef]
67. Verhaeghe, G.; Hilton, P. The effect of spot size and laser beam quality on welding performance when using high-power continuous wave solid-state lasers. *Int. Congr. Appl. Lasers Electro-Op. Laser Instit. Am.* **2005**, *2005*, 507.
68. Chowdhury, S.; Yadaiah, N.; Prakash, C.; Ramakrishna, S.; Dixit, S.; Gulta, L.R.; Buddhi, D. Laser powder bed fusion: A state-of-the-art review of the technology, materials, properties & defects, and numerical modeling. *J. Mater. Res. Technol.* **2022**, *20*, 2109–2172.
69. Mudge, R.P.; Wald, N.R. Laser-engineered net shaping advances additive manufacturing and repair. *Weld. J.-N. Y.* **2007**, *86*, 44.
70. Bi, G.; Sun, C.N.; Gasser, A. Study on influential factors for process monitoring and control in laser aided additive manufacturing. *J. Mater. Process. Technol.* **2013**, *213*, 463–468. [CrossRef]
71. Aboulkhair, N.T.; Everitt, N.M.; Ashcroft, I.; Tuck, C. Reducing porosity in AlSi10Mg parts processed by selective laser melting. *Addit. Manuf.* **2014**, *1*, 77–86. [CrossRef]
72. Chia, H.Y.; Wu, J.; Wang, X.; Yan, W. Process parameter optimization of metal additive manufacturing: A review and outlook. *J. Mater. Inf.* **2022**, *2*, 16. [CrossRef]
73. Wu, J.; Wang, X.Q.; Wang, W.; Attallah, M.M.; Loretto, M.H. Microstructure and strength of selective laser melted AlSi10Mg. *Acta Mater.* **2016**, *117*, 311–320. [CrossRef]
74. Siddique, S.; Imran, M.; Wycisk, E.; Emmelmann, C.; Walther, F. Influence of process-induced microstructure and imperfections on mechanical properties of AlSi12 processed by selective laser melting. *J. Mater. Process. Technol.* **2015**, *221*, 205–213. [CrossRef]
75. Shi, X.; Ma, S.; Liu, C.; Wu, Q.; Lu, J.; Liu, Y. Materials Science & Engineering A Selective laser melting-wire arc additive manufacturing hybrid fabrication of Ti-6Al-4V alloy: Microstructure and mechanical properties. *Mater. Sci. Eng. A* **2017**, *684*, 196–204.
76. Pérez-Ruiz, J.D.; de Lacalle, L.N.L.; Urbikain, G.; Pereira, O.; Martínez, S.; Bris, J. On the relationship between cutting forces and anisotropy features in the milling of LPBF Inconel 718 for near net shape parts. *Int. J. Mach. Tools Manuf.* **2021**, *170*, 103801. [CrossRef]
77. Pérez-Ruiz, J.D.; Martin, F.; Martínez, S.; Lamikiz, A.; Urbaikain, G.; López de Lacalle, L.N. Stiffening near-net-shape functional parts of Inconel 718 LPBF considering material anisotropy and subsequent machining issues. *Mech. Syst. Process.* **2022**, *168*, 108675. [CrossRef]
78. Trelewicz, J.R.; Halada, G.P.; Donaldson, O.K.; Manogharan, G. Microstructure and corrosion resistance of laser additively manufactured 316L stainless steel. *JOM* **2016**, *68*, 850–859. [CrossRef]
79. Wen, X.; Wan, M.; Huang, C.; Tan, Y.; Lei, M.; Liang, Y.; Cai, X. Effect of microstructure on tensile properties, impact toughness and fracture toughness of TC21 alloy. *Mater. Des.* **2019**, *180*, 107898. [CrossRef]
80. Toribio, J.; González, B.; Matos, J.C.; Ayaso, F.J. Influence of microstructure on strength and ductility in fully pearlitic steels. *Metals* **2016**, *6*, 318. [CrossRef]
81. Yang, P.; Deibler, L.A.; Bradley, D.R.; Stefan, D.K.; Carroll, J.D. Microstructure evolution and thermal properties of an additively manufactured, solution treatable AlSi10Mg part. *J. Mater. Res.* **2018**, *33*, 4040–4052. [CrossRef]
82. Koike, M.; Greer, P.; Owen, K.; Lilly, G.; Murr, L.E.; Gaytan, S.M.; Martinez, E.; Okabe, T. Evaluation of titanium alloys fabricated using rapid prototyping technologies electron beam melting and laser beam melting. *Materials* **2011**, *4*, 1776–1792. [CrossRef]
83. Hibino, S.; Todo, T.; Ishimoto, T.; Gokcekaya, O.; Koizumi, Y.; Igashira, K.; Nakano, T. Control of crystallographic texture and mechanical properties of Hastelloy-X via laser powder bed fusion. *Crystals* **2021**, *11*, 1064. [CrossRef]
84. Ishimoto, T.; Nakano, T. Microstructural Control and Functional Enhancement of Light Metal Materials via Metal Additive Manufacturing. *Mater. Trans.* **2023**, *64*, 10–16. [CrossRef]
85. Todo, T.; Ishimoto, T.; Gokcekaya, O.; Oh, J.; Nakano, T. Single crystalline-like crystallographic texture formation of pure tungsten through laser powder bed fusion. *Scr. Mater.* **2022**, *206*, 114252. [CrossRef]
86. Niendorf, T.; Leuders, S.; Riemer, A.; Richard, H.A.; Tröster, T.; Schwarze, D. Highly anisotropic steel processed by selective laser melting. *Metall. Mater. Trans. B* **2013**, *44*, 794–796. [CrossRef]
87. Nguyen, Q.B.; Luu, D.N.; Nai, S.M.; Zhu, Z.; Chen, Z.; Wei, J. The role of powder layer thickness on the quality of SLM printed parts. *Arch. Civ. Mech. Eng.* **2018**, *18*, 948–955. [CrossRef]
88. Li, J.; Ren, H.; Liu, C.; Shang, S. The effect of specific energy density on microstructure and corrosion resistance of CoCrMo alloy fabricated by laser metal deposition. *Materials* **2019**, *12*, 1321. [CrossRef]

89. Ge, W.; Guo, C.; Lin, F. Effect of process parameters on microstructure of TiAl alloy produced by electron beam selective melting. *Procedia Eng.* **2014**, *81*, 1192–1197. [CrossRef]
90. Calleja-Ochoa, A.; Gonzalez-Barrio, H.; López de Lacalle, N.; Martínez, S.; Albizuri, J.; Lamikiz, A. A new approach in the design of microstructured ultralight components to achieve maximum functional performance. *Materials* **2021**, *14*, 1588. [CrossRef]
91. Li, Y.; Yang, H.; Lin, X.; Huang, W.; Li, J.; Zhou, Y. The influences of processing parameters on forming characterizations during laser rapid forming. *Mater. Sci. Eng. A* **2003**, *360*, 18–25. [CrossRef]
92. Rott, S.; Ladewig, A.; Friedberger, K.; Casper, J.; Full, M.; Schleifenbaum, J.H. Surface roughness in laser powder bed fusion–Interdependency of surface orientation and laser incidence. *Addit. Manuf.* **2020**, *36*, 101437. [CrossRef]
93. Spierings, A.B.; Voegtlin, M.; Bauer, T.U.; Wegener, K. Powder flowability characterization methodology for powder-bed-based metal additive manufacturing. *Progress Addit. Manuf.* **2016**, *1*, 9–20. [CrossRef]
94. Abd-Elghany, K.; Bourell, D.L. Property evaluation of 304L stainless steel fabricated by selective laser melting. *Rapid Prototyp. J.* **2012**, *18*, 420–428. [CrossRef]
95. Delfs, P.; Tows, M.; Schmid, H.J. Optimized build orientation of additive manufactured parts for improved surface quality and build time. *Addit. Manuf.* **2016**, *12*, 314–320. [CrossRef]
96. Di Angelo, L.; Di Stefano, P.; Guardiani, E. Search for the optimal build direction in additive manufacturing technologies: A review. *J. Manuf. Mater. Process.* **2020**, *4*, 71. [CrossRef]
97. Calignano, F.; Lorusso, M.; Pakkanen, J.; Trevisan, F.; Ambrosio, E.P.; Manfredi, D.; Fino, P. Investigation of accuracy and dimensional limits of part produced in aluminum alloy by selective laser melting. *Int. J. Adv. Manuf. Technol.* **2017**, *88*, 451–458. [CrossRef]
98. Galy, C.; Le Guen, E.; Lacoste, E.; Arvieu, C. Main defects observed in aluminum alloy parts produced by SLM: From causes to consequences. *Addit. Manuf.* **2018**, *22*, 165–175. [CrossRef]
99. Qiu, C.; Panwisawas, C.; Ward, M.; Basoalto, H.C.; Brooks, J.W.; Attallah, M.M. On the role of melt flow into the surface structure and porosity development during selective laser melting. *Acta Mater.* **2015**, *96*, 72–79. [CrossRef]
100. Tang, M.; Pistorius, P.C. Oxides, porosity, and fatigue performance of AlSi10Mg parts produced by selective laser melting. *Int. J. Fatigue* **2017**, *94*, 192–201. [CrossRef]
101. Peng, T.; Chen, C. Influence of energy density on energy demand and porosity of 316L stainless steel fabricated by selective laser melting. *Int. J. Precis. Eng. Manuf.-Green Tech.* **2018**, *5*, 55–62. [CrossRef]
102. Coro, A.; Macareno, L.M.; Aguirrebeitia, J.; López de Lacalle, N. A methodology to evaluate the reliability impact of the replacement of welded components by additive manufacturing spare parts. *Metals* **2019**, *9*, 932. [CrossRef]
103. Wei, C.; Li, L. Recent progress and scientific challenges in multi-material additive manufacturing via laser-based powder bed fusion. *Virtual Phys. Prototyp.* **2021**, *16*, 347–371. [CrossRef]
104. Kaufmann, N.; Imran, M.; Wischeropp, T.M.; Emmelmann, C.; Siddique, S.; Walther, F. Influence of process parameters on the quality of aluminum alloy EN AW 7075 using selective laser melting (SLM). *Phys. Procedia* **2016**, *83*, 918–926. [CrossRef]
105. Sames, W.J.; List, F.A.; Pannala, S.; Dehoff, R.R.; Babu, S.S. The metallurgy and processing science of metal additive manufacturing. *Int. Mater. Rev.* **2016**, *61*, 315–360. [CrossRef]
106. Parthasarathy, J.; Starly, B.; Raman, S.; Christensen, A. Mechanical evaluation of porous titanium (Ti6Al4V) structures with electron beam melting (EBM). *J. Mech. Behav. Biomed. Mater.* **2010**, *3*, 249–259. [CrossRef]
107. Kimura, T.; Nakamoto, T. Microstructures and mechanical properties of A356 (AlSi7Mg0. 3) aluminum alloy fabricated by selective laser melting. *Mater. Des.* **2016**, *89*, 1294–1301. [CrossRef]
108. Gong, H.; Rafi, K.; Gu, H.; Ram, G.J.; Starr, T.; Stucker, B. Influence of defects on mechanical properties of Ti–6Al–4 V components produced by selective laser melting and electron beam melting. *Mater. Design* **2015**, *86*, 545–554. [CrossRef]
109. Tan, H.; Wong, E.; Dalgarno, W. An overview of powder granulometry on feedstock and part performance in the selective laser melting process. *Addit. Manu.* **2017**, *18*, 228–255. [CrossRef]
110. Meier, H.; Haberland, C. Experimental studies on selective laser melting of metallic parts. *Mat.-Wiss. U. Werkst.* **2008**, *39*, 665–670. [CrossRef]
111. Mercelis, P.; Kruth, J.P. Residual stresses in Selective Laser Sintering and Selective Laser Melting. *Rapid Prototyp. J.* **2005**, *12*, 109–131. [CrossRef]
112. Yadroitsev, I.; Matope, S.; Dimitrov, D.; Mugwagwa, L. Investigation of the effect of scan vector length on residual stresses in selective laser melting of maraging steel 300. *S. Afr. J. Ind. Eng.* **2019**, *30*, 60–70.
113. Zaeh, M.F.; Branner, G. Investigations on residual stresses and deformations in selective laser melting. *Prod. Eng.* **2010**, *4*, 35–45. [CrossRef]
114. Kempen, K.; Vrancken, B.; Thijs, L.; Buls, S.; Van, H.J.; Kruth, J.P. Lowering thermal gradients in selective laser melting by pre-heating the baseplate. In Proceedings of the Solid Freeform Fabrication Symposium Proceedings, Austin, TX, USA, 12–15 August 2023; pp. 12–14.
115. Mugwagwa, L.; Yadroitsev, I.; Matope, S. Effect of process parameters on residual stresses, distortions, and porosity in selective laser melting of maraging steel 300. *Metals* **2019**, *9*, 1042. [CrossRef]
116. Thomas, B.R.; Bibb, R. An investigation into the geometric constraints of selective laser melting for the development of design rules. In Proceedings of the 9th National Conference on Rapid Design, Prototyping & Manufacture, High Wycombe, UK, 13 June 2008; pp. 11–20.

117. Yadroitsev, I.; Yadroitsava, I.; Bertrand, P.; Smurov, I. Factor analysis of selective laser melting process parameters and geometrical characteristics of synthesized single tracks. *Rapid Prototyp. J.* **2012**, *12*, 201–208. [CrossRef]
118. Thijs, L.; Verhaeghe, F.; Craeghs, T.; Van, H.J.; Kruth, J.P. A study of the microstructural evolution during selective laser melting of Ti–6Al–4V. *Acta Mater.* **2010**, *58*, 3303–3312. [CrossRef]
119. Gu, H.; Gong, H.; Pal, D.; Rafi, K.; Starr, T.; Stucker, B. Influences of energy density on porosity and microstructure of selective laser melted 17–4PH stainless steel. In *Proceedings of the Annual International Solid Freeform Fabrication Symposium*; University of Texas at Austin: Austin, TX, USA, 2013.
120. Delgado, J.; Ciurana, J.; Rodríguez, C.A. Influence of process parameters on part quality and mechanical properties for DMLS and SLM with iron-based materials. *Int. J. Adv. Manuf. Technol.* **2012**, *60*, 601–610. [CrossRef]
121. Pauzon, C.; Mishurova, T.; Evsevleev, S.; Dubiez-Le Goff, S.; Murugesan, S.; Bruno, G.; Hryha, E. Residual stresses and porosity in Ti-6Al-4V produced by laser powder bed fusion as a function of process atmosphere and component design. *Addit. Manuf.* **2021**, *47*, 102340. [CrossRef]
122. Kruth, J.P.; Deckers, J.; Yasa, E.; Wauthlé, R. Assessing and comparing influencing factors of residual stresses in selective laser melting using a novel analysis method. Proceedings of the institution of mechanical engineers. *Part B J. Eng. Manuf.* **2012**, *226*, 980–991. [CrossRef]
123. Casavola, C.; Campanelli, S.L.; Pappalettere, C. Preliminary investigation on distribution of residual stress generated by the selective laser melting process. *J. Strain Anal. Eng. Des.* **2009**, *44*, 93–104. [CrossRef]
124. Weller, C.; Kleer, R.; Piller, F.T. Economic Implications of 3D Printing. *Int. J. Prod. Econ.* **2015**, *3*, 303–312.
125. Franchitti, S.; Borrelli, R.; Pirozzi, C.; Carrino, L.; Polini, W.; Sorrentino, L.; Gazzerro, A. Investigation on Electron Beam Melting: Dimensional accuracy and process repeatability. *Vacuum* **2018**, *157*, 340–348. [CrossRef]
126. Abedi, R.; Hanzaki, Z.; Azami, M.; Kahnooji, M.; Rahmatabadi, D. The high temperature flow behavior of additively manufactured Inconel 625 superalloy. *Mater. Res. Express* **2019**, *6*, 116514. [CrossRef]

metals

MDPI

Article

Effect of Co-Coated Al_2O_3 Composite Powders on LPBF-Manufactured CoCrMo Alloy

Lehui Zhang [1,2], Wei Liu [1,2,*], Yu Cao [1,2], Haoxin Sun [1,2], Qiusheng Xie [1,2], Junyun Lai [1,2] and Peikang Bai [1,2]

1 School of Materials Science and Engineering, North University of China, Taiyuan 030051, China
2 Shanxi Key Laboratory of Controlled Metal Solidification and Precision Manufacturing, Taiyuan 030051, China
* Correspondence: lwnuc@163.com; Tel.: +86-351-3559638

Abstract: In this study, we systematically examined the influence mechanisms of introduced cobalt coated alumina (Co-coated Al_2O_3) particles on the microstructure and properties of cobalt-chromium-molybdenum (CoCrMo) alloy printed by Laser Powder Bed Fusion (LPBF). The Co-coated Al_2O_3 composite powders with different density of cobalt coating were prepared by varying Al_2O_3 load from 1 g/150 mL to 2 g/150 mL during the electroless plating process. Then they were mixed with CoCrMo powders in the proportion of 1 wt.% and formed standard samples by LPBF technology. The results showed that the addition of Co-coated Al_2O_3 particles improved the friction performance of CoCrMo alloys significantly. The wear depth of CCM@2Al_2O_3 was only 2.18 μm and the wear volume of it was about 10% of pure CoCrMo alloy. The CoCrMo alloy introduced the Co-coated Al_2O_3 particles with a 1 g/150 mL Al_2O_3 load formed metal-ceramic bonding interface, which solved the problem of poor wettability between Al_2O_3 and matrix in LPBF process. Such CoCrMo alloy exhibited excellent tensile properties and the mean microhardness of it reached 379.9 ± 3.5 $HV_{0.5}$.

Keywords: alumina; cobalt-chromium-molybdenum alloy; electroless plating; friction performance; implants; Laser Powder Bed Fusion

Citation: Zhang, L.; Liu, W.; Cao, Y.; Sun, H.; Xie, Q.; Lai, J.; Bai, P. Effect of Co-Coated Al_2O_3 Composite Powders on LPBF-Manufactured CoCrMo Alloy. *Metals* **2023**, *13*, 310. https://doi.org/10.3390/met13020310

Academic Editors: Eric Hug and Reza Ghomashchi

Received: 24 November 2022
Revised: 26 January 2023
Accepted: 31 January 2023
Published: 3 February 2023

1. Introduction

In recent years, Additive Manufacturing (AM) has attracted significant attention as a type of novel and continuously expanding technology [1,2]. Laser Powder Bed Fusion (LPBF), an additive manufacturing process, is widely used to manufacture orthopedic implants [3,4]. The combination of CAD technology in LPBF makes the implants more accurate and easier to meet the requirements of personalized medicine [5–9]. It is generally known that the most-used metallic biomaterials in LPBF-manufactured implants are titanium alloys, Cobalt-chromium (CoCr) alloys, and stainless steel [10,11]. And CoCr alloys plays an irreplaceable role in orthopedic implants due to excellent mechanical properties and biocompatibility [12–16]. However, the service time of the artificial knee joints is limited because of the abrasion during movement, and the second operation is often required [17]. Additionally, more young and active patients are using artificial knee joints, and they have higher requirements for the wear resistance of implants [18].

Therefore, a great focus of researches have been put into improving the wear resistance of LPBF-processed CoCr alloys. Li et al. [19] carried out three kinds of heat treatments on CoCrMo alloys respectively, finding that the friction performance of aging-treated alloys at moderate temperatures (450 °C to 750 °C) had been improved. Balagna et al. [20] prepared the coating of tantalum carbide on the surface of CoCrMo alloys by molten salts, and the existence of the coating reduced the friction coefficient. However, this method did not control the thickness of the coating and was not suitable for personalized medicine. Isik et al. [21] used two premix raw powders (CoCrMo + 2 wt.% tricalcium phosphate and CoCrMo + 2 wt.% tricalcium phosphate + 4 wt.% Al_2O_3) to surface coat the CoCr alloy via Laser Engineered Net Shaping. This method improved the wear resistance of CoCr alloys, but the inherent limitation of coatings was that they introduced discontinuities in mechanical properties at the interface of the alloy matrix.

Some scholars proposed to enhance the wear resistance of LPBF-processed CoCr alloys by improving the properties ofpowder feedstocks. Liu et al. [22] pointed out that it was a very effective method to improve the friction performance by adding high wear resistance powder to the matrix powder for parts fabricated by LPBF technology. The bioceramics such as Al_2O_3, zirconium dioxide (ZrO_2) and silicon nitride (Si_3N_4) are usually used for the modification of metallic biomaterials [23–25]. Among them, Al_2O_3 is the preferred ceramicbecause of its excellent mechanical properties and wear resistance [26–30]. Gu et al. [31] prepared Al-based composites reinforced by Al_2O_3 particles using LPBF technology, which exhibited superior friction performance. However, insufficient wettability between Al and Al_2O_3 led to plenty of defects such as interfacial microcracks and interface separation [32]. Li et al. [33] proposed the method of producing MMC powder through electroless plating, successfully preparing Ni/Al_2O_3 composite powder and the composite powder was suitable for LPBF forming. Cao et al. [34] prepared Ni-Al_2O_3 particles with a core-shell structure by electroless plating, and successfully fabricating CoCrMo alloys with different amounts of Ni-Al_2O_3 particles by LPBF process. They found that the increase of Ni-Al_2O_3 content helped to improve the wear resistance of the alloys. The composite containing 1 wt.% Ni-Al_2O_3 particles achieved the highest microhardness and the most stable coefficient of friction. However, the effect of Ni-Al_2O_3 powder prepared by different electroless plating processes on the properties of composite was not explored. Moreover, the introduction of Ni-Al_2O_3 particles increased nickel content in the CoCrMo alloys, which made the implants easier to release Ni ions during wear and tear. Nickel was a harmful element to humans because it could cause a biological allergy [35].

In order to improve the application of modified CoCrMo alloys in orthopedic implants, Al_2O_3 powder was coated with nontoxic cobalt layer using the electroless plating process. Moreover, the melting point of cobalt is close to that of CoCrMo alloy, so the cobalt layer helped the wetting of Al_2O_3 particles by matrix alloy. Importantly, the effect of Co-coated Al_2O_3 composite powders with different coating quality on the microstructure and properties of these CoCrMo alloys were explored.

2. Materials and Methods

2.1. Materials

Commercially available CoCrMo powders produced by Guangzhou Nalian Co., Ltd. (Guangzhou, China) were used in this study, and their chemical composition is listed in Table 1. The SEM micrograph, Figure 1a, shows the spherical morphology of the CoCrMo powder particles having uniform size. The average size of CoCrMo powders was 26 μm. The Al_2O_3 powders used in electroless plating were produced by Jiyuan Ceramic Material Co., Ltd. (Jiyuan, China). The smooth surface of as-received Al_2O_3 (Figure 1b) was conducive to coating the cobalt layer during electroless plating process. And Figure 2 shows the schematic representation of electroless Co plating and LPBF process.

Table 1. Chemical composition of CoCrMo powder (wt.%).

Co	Cr	Mo	Si	Fe	N	O	C
66.53	26.81	5.72	0.69	0.11	<0.1	<0.1	<0.01

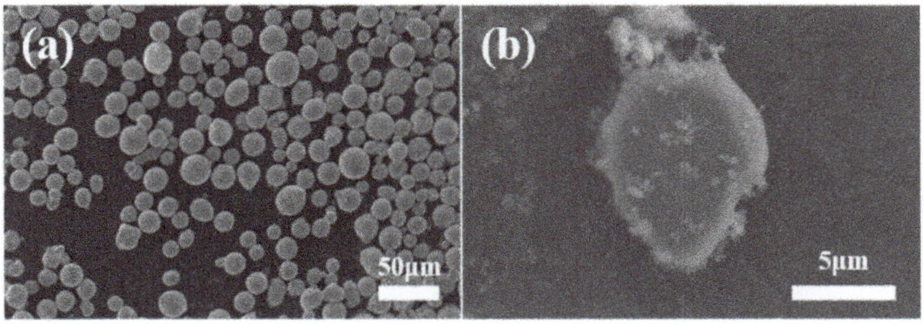

Figure 1. SEM images of powders: (**a**) CoCrMo powder; (**b**) Al_2O_3 powder.

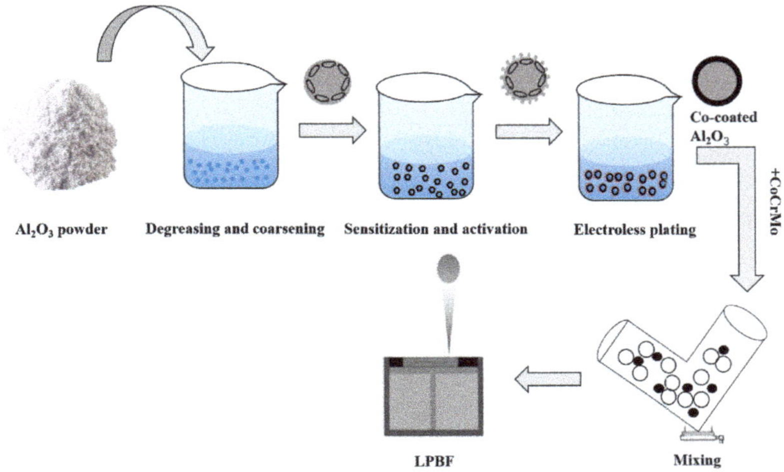

Figure 2. Schematic representation of electroless Co plating and LPBF process.

2.2. Fabrication Procedure

2.2.1. Electroless Plating

In the process of electroless plating, the Al_2O_3 powders were pretreated before putting into the plating solution. First of all, the raw powders were placed in 5 mol/L NaOH solution for degreasing, and then coarsened with HCl solution. The coarsened Al_2O_3 powders needed to be sensitized and activated successively. The sensitization solution was $SnCl_2$ solution (20 g/L), and the concentration of $PdCl_2$ activation solution was 0.5 g/L. The above four steps required magnetic stirring of the solution for 20 min.

The pretreated Al_2O_3 powders were placed in a plating beaker immersed in a 66 °C water bath and stirred for 30 min. The composition of the plating solution is listed in Table 2, consisting of the main salt ($CoSO_4 \cdot 7H_2O$), reducing agent ($N_2H_4 \cdot H_2O$), stabilizer agent ($C_4H_6O_6$), complexing agent (EDTA), and pH regulating agent (NaOH). These chemicals were produced by Chengdu Colon Chemicals Co., Ltd. (Chengdu, China). This process was mainly the reaction of reducing agent and cobalt salt, and the equation of chemical reaction was shown in Equation (1). The plating solution and Al_2O_3 powders were separated by centrifugation, and subsequently the powders were dried in a vacuum for 2 h to get Co-coated Al_2O_3 composite powders.

$$2Co^{2+} + N_2H_4 + 4OH^- \rightarrow 2Co + N_2\uparrow + 4H_2O \quad (1)$$

Table 2. Chemical composition of the plating solution.

$CoSO_4 \cdot 7H_2O$ (g/L)	$N_2H_4 \cdot H_2O$ (ml/L)	$C_4H_6O_6$ (g/L)	EDTA (g/L)	NaOH (mol/L)
35	30	10	8	5

The powder load had an important influence on the properties of the coating during electroless plating [36]. There were more powder particles in the same volume of plating solution when the powder load was large, so the surface coating density of a single powder particle decreased. In this study, the powder load was expressed by the ratio of the Al_2O_3 weight to the volume of the plating solution. Co-coated Al_2O_3 composite powders with Al_2O_3 load of 1 g/150 mL, 1.5 g/150 mL and 2 g/150 mL were prepared respectively. The amount of Al_2O_3 added to CoCrMo alloy was selected as 1 wt.% according to previous studies [34], so the Co-coated Al_2O_3 composite powders were mixed with CoCrMo powders in a mass ratio of 1:99 by V-type mixer (WKA-100, Qingzhou Madsen Co., Ltd., Qingzhou, China). The mixer ran for 6 h at a speed of 200 RPM.

2.2.2. LPBF process

The pure CoCrMo alloy (CCM) and the CoCrMo alloys containing Co-coated Al_2O_3 particles were prepared by LPBF on an E-PLUS 150 machine (Beijing E-Plus 3D Tech. Co., Ltd., Beijing, China), and the process was carried out on the 316 L stainless steel baseplate at the deposition conditions in Table 3. 10 mm × 10 mm × 10 mm blocks (Figure 3a) were fabricated to analyze the microstructure and properties, which longitudinal section was parallel to the building direction. The tensile samples were formed according to the dimensions shown in Figure 3b. In this paper, the CoCrMo alloys containing Co-coated Al_2O_3 particles were named CCM@1Al_2O_3, CCM@1.5Al_2O_3 and CCM@2Al_2O_3 respectively, corresponding to the Al_2O_3 load of 1 g/150 mL, 1.5 g/150 mL and 2 g/150 mL during electroless plating process (these names were used in the following discussion).

Table 3. LPBF process parameters.

Laser Power	Scan Speed	Layer Thickness	Scan Spacing
120 W	760 mm/s	30 μm	80 μm

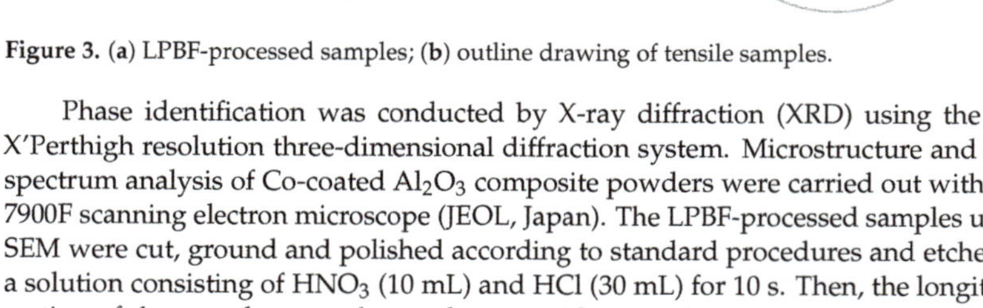

Figure 3. (**a**) LPBF-processed samples; (**b**) outline drawing of tensile samples.

Phase identification was conducted by X-ray diffraction (XRD) using the Dutch X'Perthigh resolution three-dimensional diffraction system. Microstructure and energy spectrum analysis of Co-coated Al_2O_3 composite powders were carried out with a JSM-7900F scanning electron microscope (JEOL, Japan). The LPBF-processed samples used for SEM were cut, ground and polished according to standard procedures and etched with a solution consisting of HNO_3 (10 mL) and HCl (30 mL) for 10 s. Then, the longitudinal section of the samples was observed using a JSM-7900F scanning electron microscope. Prior to EBSD analysis, argon ion etching of the metallography surfaces was conducted to provide stress- free surfaces [37]. The Vickers hardness of samples was measured using a microhardness tester (HV-1000B, Lanzhou Huayin Test Instrument Co., Ltd., Lanzhou, China) at a load of 4.9 N and an indentation time of 10 s. All samples were measured at 10 points in the cross section and final microhardness value was the average of these data. The tensile test was conducted at room temperature on a 3382 universal material testing machine (Instron, Norwood, MA, USA) at a tensile speed of 0.5 mm/min. The test process needed to measure the gauge length, thickness and width of the samples, and then clamped the samples to the machine. The friction and wear test were conducted with the HSR-2M reciprocating dry friction and wear testing machine (Lanzhou Zhongke Kaihua Technology Development Co., Ltd., Lanzhou, China) with a 3 mm silicon nitride friction pair. The machine applied a load of 5 N for 30 min at a speed of 500 RPM. Then the wear depth and wear volume of the samples were measured by the MT-500 probe type material surface abrasion tester (Lanzhou Zhongke Kaihua Technology Development Co., Ltd., Lanzhou, China).

3. Results

3.1. Co-Coated Al_2O_3 Composite Powder

3.1.1. Phase Analysis

Figure 4 shows XRD patterns of Co-coated Al_2O_3 composite powders. The main characteristic peaks were Al_2O_3 and Co in the patterns. The peaks of Co were detected at 41.68°, 44.76° and 47.57° respectively, which proved that Co was generated during electroless plating successfully. It was found that the diffraction peak height of Co was close to that of Al_2O_3 in the composite powder with 1 g/150 mL Al_2O_3 load, indicating that the Co phase accounted for a large proportion in the composite powder. However, the intensity of Co peak gradually decreased with an increase of Al_2O_3 load. Because the Al_2O_3 load went up, there were more Al_2O_3 particles in the same volume of plating solution, resulting in less Co deposited on each Al_2O_3 particle. Notably, the peak of Co_3O_4 appeared at 36.84° in the patterns. The reason was that a small amount of cobalt salt generated into $Co(OH)_2$ under the joint action of the reducing agent and the complexing agent in the plating solution. $Co(OH)_2$ decomposed into Co_3O_4 under the condition of rapid heat release during electroless plating process [38,39].

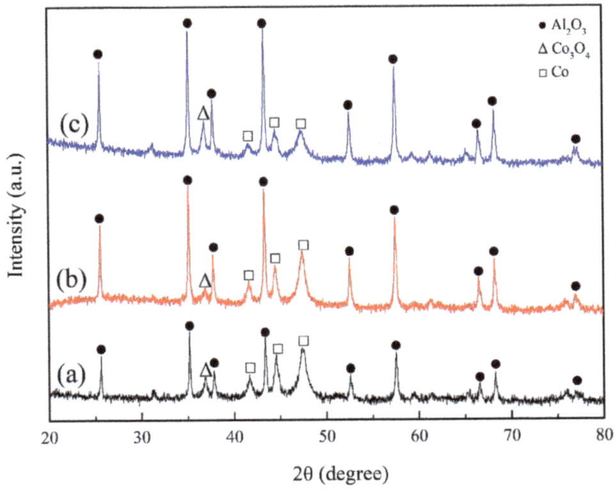

Figure 4. XRD patterns of Co-coated Al_2O_3 composite powders prepared by different Al_2O_3 load: (**a**) 1 g/150 mL, (**b**) 1.5 g/150 mL and (**c**) 2 g/150 mL.

3.1.2. Morphology Analysis

Figure 5 illustrates the morphological images of Co-coated Al_2O_3 composite powders. As shown in the low magnification maps (Figure 5(a1–c1)), the surface of the composite powders showed the flocculent coating compared with the raw Al_2O_3 powders, and all particles were coated successfully. The specific powder particles were selected for further observation (Figure 5(a2–c2)). The composite powder with Al_2O_3 load of 1 g/150 mL also had excess cobalt in addition to forming a cobalt coating, which was different from the other two powders. The cobalt coatings of different Co-coated Al_2O_3 particles were observed at higher magnification (Figure 5(a3–c3)). It was found that the Co-coated Al_2O_3 particle with 1 g/150 mL Al_2O_3 load presented a dense cobalt coating, while the cobalt coatings of the other two powders had pores. The porosity of cobalt coating increased with the enhancement of Al_2O_3 load. The reason was that less cobalt deposited on the surface of Al_2O_3 when the composite powder was prepared with a larger Al_2O_3 load, and it led to a decrease in the density of cobalt coating.

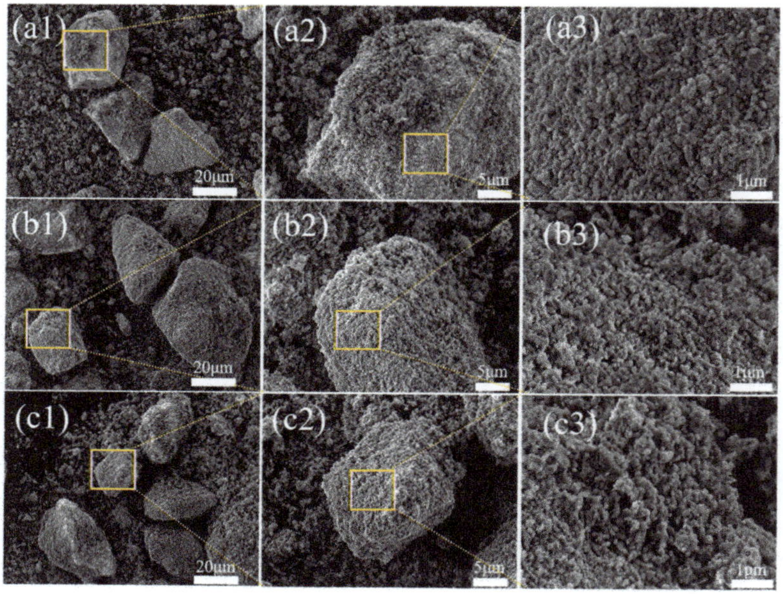

Figure 5. SEM images of Co-coated Al_2O_3 composite powders prepared by different Al_2O_3 load: (**a1–a3**)1 g/150 mL, (**b1–b3**) 1.5 g/150 mL and (**c1–c3**) 2 g/150 mL.

Figure 6 shows the EDS results of Co-coated Al_2O_3 composite powders. It was found that the main elements detected in the powders were O, Al and Co, which proved that the flocculent coating of the composite powders was Co. The cobalt coating of Co-coated Al_2O_3 particle with 1 g/150 mL Al_2O_3 load was completely detected. And the content of Co decreased with the increase of Al_2O_3 load. The Co-coated Al_2O_3 particle with 2 g/150 mL Al_2O_3 load even had a partially missing of cobalt coating, because its cobalt coating was the loosest.

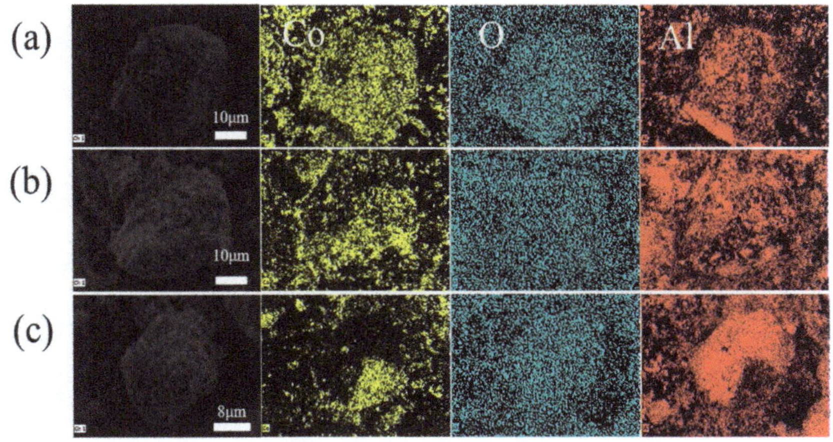

Figure 6. EDS of Co-coated Al_2O_3 composite powders prepared by different Al_2O_3 load: (**a**) 1 g/150 mL, (**b**) 1.5 g/150 mL and (**c**) 2 g/150 mL.

3.2. LPBF-Processed Parts

3.2.1. Microstructural Morphology

Figure 7 shows the cross-section microstructure of four CoCrMo alloys fabricated by LPBF. By observing the low magnification maps (Figure 7(a1–d1)), it was found that all samples were dense and no obvious pores appeared. The number of Co-coated Al_2O_3 particles increased in CoCrMo alloys with the enhancement of Al_2O_3 load. As can be seen from (Figure 7(a2–d2)), CoCrMo alloys exhibited a typical fish scale melting trajectory, and the lamellar semi-ellipse observed in the figure was the boundary of the molten pool [40]. Such molten pool morphology is typical of the longitudinal section for LPBF-processed parts [41]. All molten pools in the corresponding figure were measured and final molten pool depth of the alloy was the average of these data. The results showed that the molten pool depth of CCM, CCM@1Al_2O_3, CCM@1.5Al_2O_3 and CCM@2Al_2O_3 were 47.24 μm, 35.24 μm, 34.88 μm and 32.68 μm respectively. It was found that molten pool of the CoCrMo alloys containing Co-coated Al_2O_3 particles was shallower than CCM. Because the temperature exceeded 1500 °C during LPBF process [42]. The laser acted on the Al_2O_3 after melting the cobalt coating when it scanned the Co-coated Al_2O_3 particles. However, the absorptivity of Al_2O_3 is only 0.173 [43], which is far lower than that of CoCrMo [42]. The formed powders containing the Co-coated Al_2O_3 particles got less energy inLPBF process, so the molten pool depth of these CoCrMo alloys decreased.

Figure 7(a3–d3) presents the SEM images at higher magnification. It was found that the LPBF-processed parts showed conventional columnar and equiaxial dendrites. The metal-ceramic bonding interface was formed in CCM@1Al_2O_3, which may have a favorable effect on the alloy. But no similar phenomenon was observed in CCM@1.5Al_2O_3 and CCM@2Al_2O_3, there were defects around the Al_2O_3 particles. Because thicker Co-coating of Al_2O_3 particles have helped particles wettability by the matrix alloy while when the Co content (thinner coatings) was low, the Al_2O_3 particles could not form strong bond with the matrix.

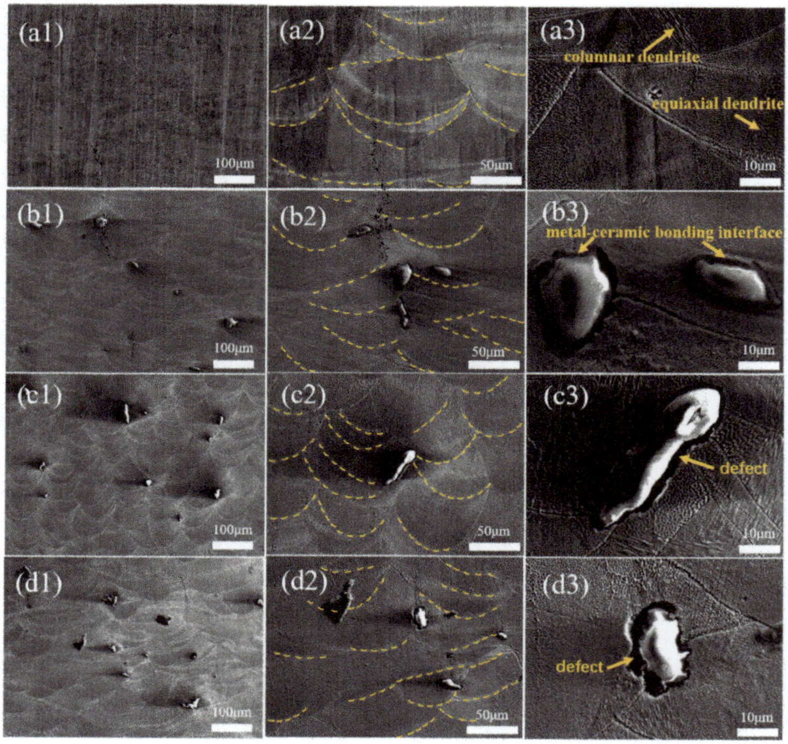

Figure 7. SEM images of longitudinal section for the LPBF-processed parts: (**a1–a3**) CCM, (**b1–b3**) CCM@1Al_2O_3, (**c1–c3**) CCM@1.5Al_2O_3 and (**d_1–d_3**) CCM@2Al_2O_3.

3.2.2. Phase Analysis

Figure 8 depicts the XRD patterns of the LPBF-processed parts obtained over a wide range of 2θ (40–100°). The diffraction peaks of ε(hcp)CoCrMo and γ(fcc)CoCrMo were identified clearly. However, the CoCrMo alloys containing Co-coated Al_2O_3 particles did not show the diffraction peaks of Al_2O_3, as the amount of Co-coated Al_2O_3 particles was only 1 wt.% in alloys. This phenomenon also occurred in other metal-matrix ceramic composites [44]. By comparing the relative intensity of peaks, it was found that the γ peaks strength of the CoCrMo alloys containing Co-coated Al_2O_3 particles were lower than that of CCM. The reason was that the addition of Al_2O_3 reduced the cooling rate of alloys [45]. The γ phase was stable at high temperatures, and the reduction of the cooling rate was conducive to the conversion of γ phase to ε phase [46]. The cooling rate of CCM@1Al_2O_3 was the fastest in three CoCrMo alloys containing Co-coated Al_2O_3 particles, because the metal-ceramic bonding interface played an auxiliary role in the heat conduction of Al_2O_3. And the other two alloys had relatively slow cooling rate, in which the heat transfer of Al_2O_3 depended on dissipation. The cooling rate of CCM@2Al_2O_3 was faster than that of CCM@1.5Al_2O_3, because the cobalt coating of Co-coated Al_2O_3 particles in CCM@2Al_2O_3 had more pores to dissipate heat. Finally, the intensity of the γ diffraction peaks of CCM@1.5Al_2O_3 was the weakest in them.

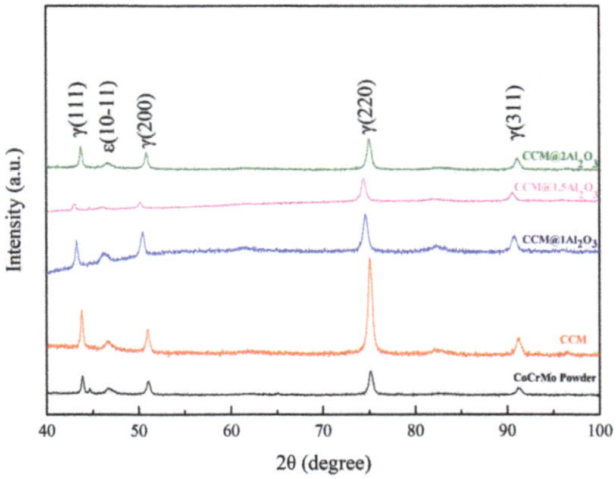

Figure 8. Phase analysis results measured by XRD.

3.2.3. Kernel Average Misorientation (KAM)

Figure 9 presents the KAM of longitudinal section for the LPBF-processed parts. The results showed that all samples retained large strain due to the very high cooling rate and repeated layer stacking during LPBF process. KAM analysis of the CoCrMo alloys containing Co-coated Al_2O_3 particles showed lower residual stresses, higher concentration of blue regions, than that of CCM, indicating that these samples had less stored energy [47]. Because they were formed at a relatively low cooling rate, more energy was consumed during the cooling process. At the same time, the KAM could reflect the dislocation density of alloys. And the KAM of CCM@1.5Al_2O_3 was the lowest by comparing the intensity of the ribbon, revealing the lowest dislocation density in this sample.

3.2.4. Friction Performance

Figure 10 shows the wear tracks of the CoCrMo alloys. It was found that all samples exhibited similar shape of the wear tracks. The maximum wear depth and wear volume obtained in the test were analyzed statistically (Figure 11). The wear depth of CCM was the largest among all samples, reaching 9.2665μm. The Co-coated Al_2O_3 particles significantly reduced the wear depth of the CoCrMo alloys. The wear depth of the CoCrMo alloys containing Co-coated Al_2O_3 particles decreased gradually with the increase of Al_2O_3 load. The wear volume was another important parameter reflecting wear resistance of materials. It was found that the wear volume showed a similar trend to the wear depth, and the wear volume of CCM@1Al_2O_3 was only about 50% of CCM. These two groups of data showed that the addition of Al_2O_3 substantially enhanced the wear resistance of CoCrMo alloy fabricated by LPBF [21]. And CCM@2Al_2O_3 showed the most excellent friction performance. Because the composite powder prepared using a large load contained a small proportion of cobalt in Co-coated Al_2O_3 particles. The density of Co is about 2.5 times larger than that of Al_2O_3, so the average density of Co-coated Al_2O_3 composite powder with a large load was low. As 1 wt.% of composite powders were mixed with CoCrMo powders for all samples, the number of Co-coated Al_2O_3 particles in CoCrMo alloys increased with the enhancement of Al_2O_3 load.

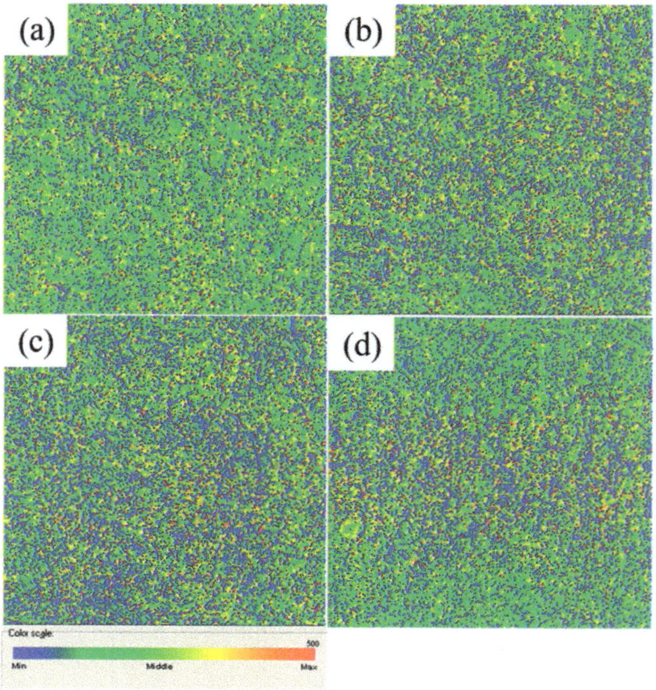

Figure 9. The KAM of longitudinal section for the LPBF-processed parts: (**a**) CCM, (**b**) CCM@1Al_2O_3, (**c**) CCM@1.5Al_2O_3 and (**d**) CCM@2Al_2O_3.

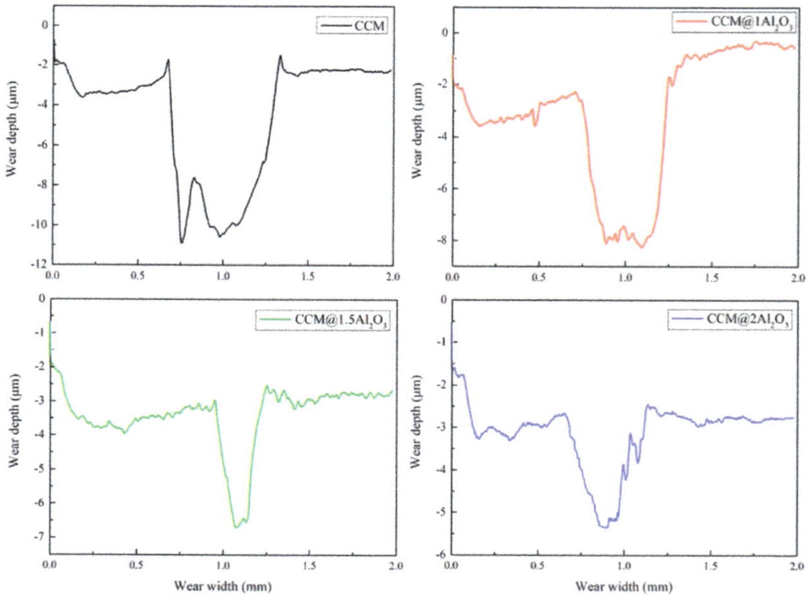

Figure 10. The wear tracks of the LPBF-processed parts.

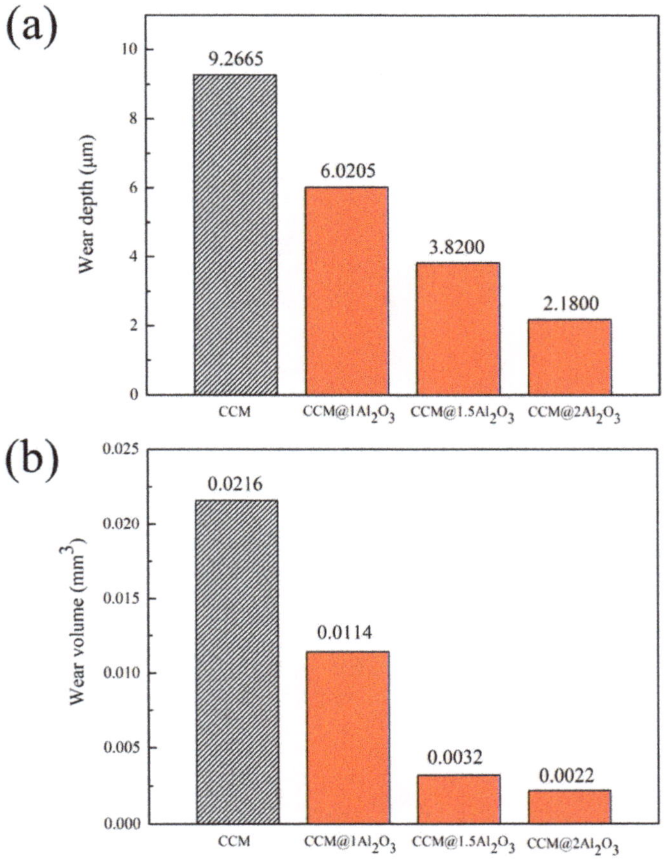

Figure 11. Friction data of the LPBF-processed parts: (**a**) wear depth; (**b**) wear volume.

3.2.5. Mechanical Property

Figure 12 shows the microhardness measured on LPBF-processed parts. The mean microhardness of CCM, CCM@1Al_2O_3, CCM@1.5Al_2O_3 and CCM@2Al_2O_3 were 358.5 ± 3.2 $HV_{0.5}$, 379.9 ± 3.5 $HV_{0.5}$, 373.1 ± 4.3 $HV_{0.5}$ and 368.7 ± 3.7 $HV_{0.5}$, respectively. The high microhardness obtained at the CoCrMo alloys containing Co-coated Al_2O_3 particles indicated that the addition of Al_2O_3 ehanced the microhardness of CoCrMo alloys. On the one hand, Al_2O_3 existed in the CoCrMo alloys as the second phase particles, which had a dispersive strengthening effect on the alloys [48]. On the other hand, CoCrMo alloys containing Co-coated Al_2O_3 particles had more ε phase, contributing to the increase in microhardness [22]. The maximum mean microhardness was obtained at CCM@1Al_2O_3, due to the metal-ceramic bonding interface making the diffusion strengthening effect more significant. Although the XRD results showed that the content of ε phase in CCM@1Al_2O_3 was not the highest among all samples, the experiment indicated that the microhardness enhancement effect brought by the metal-ceramic bonding interface was stronger than the ε phase. However, the maximum average microhardness for the Al_2O_3 modified CoCr alloys was only 338.8 HV in the study of Hong [49], because the introduced Al_2O_3 particles were not treated by electroless plating.

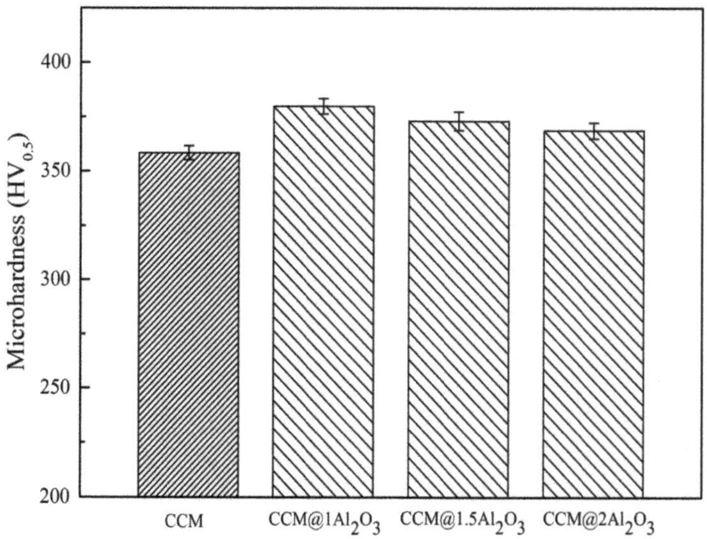

Figure 12. Microhardness of the LPBF-processed parts.

Figure 13 depicts the results of room-temperature tensile tests on the as-printed samples. The ultimate tensile strength (UTS) was directly measured by the standard tensile test, and the elongation was obtained by calculating the ratio of the displacement to the gauge length of the tensile samples (Table 4). It was found that the UTS of CoCrMo alloys containing Co-coated Al_2O_3 particles were lower than that of CCM, but all the values of UTS were higher than 665 MPa required in ISO 5832-4-2014. The decrease in strength of them was due to the relatively low dislocation density compared to CCM. It was well known that a high dislocation density can slow down the movement of the dislocation, which was beneficial to the strength. And the UTS of CCM@1.5Al_2O_3 was the minimum, because it showed the lowest dislocation density. In addition, defects also influenced the strength of composite materials [50]. The metal-ceramic bonding interface was formed in CCM@1Al_2O_3, which made it no defects around Al_2O_3 and contributed to the strength of the alloy. In terms of ductility, the increase of ε phase in CoCrMo alloys containing Co-coated Al_2O_3 particles led to the decrease of elongation, because the HCP structure of ε phase was not conducive to the crystal slip during the tensile process. The elongation of CCM@1Al_2O_3 was close to that of CCM, indicating the ductility of it was excellent.

Although comprehensive analysis showed that CCM@1Al_2O_3 were excellent in UTS and elongation, the values were still lower than that of CCM. Further researches were needed to improve the tensile properties of the LPBF-processed Al_2O_3 modified CoCrMo alloys.

Table 4. Tensile properties of the LPBF-processed parts.

Samples	UTS (MPa)	Elongation (%)
CCM	1191.46	41.01
CCM@1Al_2O_3	1089.07	37.76
CCM@1.5Al_2O_3	1050.07	25.47
CCM@2Al_2O_3	1072.93	29.46

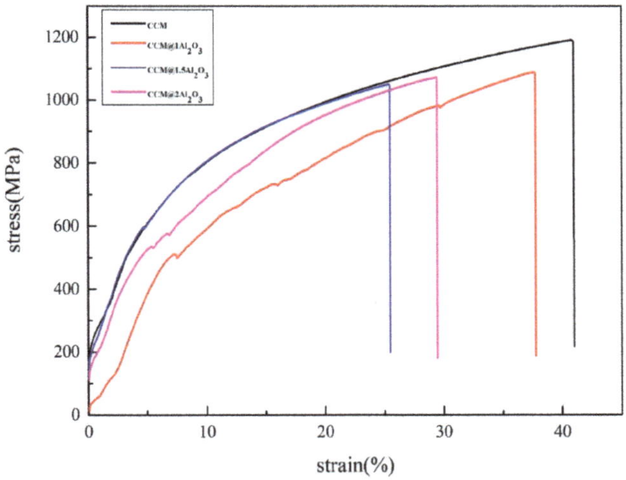

Figure 13. Stress-strain curves of the LPBF-processed parts.

4. Conclusions

To summarize, the effect of Co-coated Al_2O_3 composite powders prepared with different Al_2O_3 load on the microstructure and properties of LPBF-processed Al_2O_3 reinforced CoCrMo alloys were systematically investigated. According to SEM and EDS, the density of cobalt coating for the Co-coated Al_2O_3 composite powders decreased with the increase of the Al_2O_3 load in the electroless plating process. XRD patterns of LPBF-processed parts showed that the CoCrMo alloys containing Co-coated Al_2O_3 particles had more phase transitions from γ phase to ε phase. The ε phase was beneficial to microhardness of CoCrMo alloys but reduced the elongation of them. The wear depth of CCM@2Al_2O_3 was only 2.18 μm and the wear volume of it was about 10% of CCM. The metal-ceramic bonding interface was formed in CCM@1Al_2O_3, which helped it maintaining excellent tensile properties.The mean microhardness of CCM@1Al_2O_3 reached 379.9 $\pm$ 3.5 $HV_{0.5}$, which was about 6% higher than CCM. However, more research is needed to optimize the friction and tensile properties of CoCrMo alloy at the same time in the future.

Author Contributions: Conceptualization, L.Z. and W.L.; validation, Y.C. and H.S.; formal analysis, Q.X. and J.L.; resources, P.B. and W.L.; data curation, W.L. and Y.C.; writing—original draft preparation, L.Z.; writing—review and editing, W.L.; visualization, H.S., Q.X. and J.L.; supervision, P.B. All authors have read and agreed to the published version of the manuscript.

Funding: This research and the APC was funded by Special Project for Scientific and Technology Cooperation and Exchange of Shanxi Province (No. 202104041101010) and Primary Research and Development Plan of Shanxi Province (No. 202102050201009).

Institutional Review Board Statement: Not applicable.

Informed Consent Statement: Not applicable.

Data Availability Statement: Not applicable.

Acknowledgments: The authors are grateful for these grants.

Conflicts of Interest: The authors declare no conflict of interest.

References

1. Wang, J.H.; Ren, J.; Liu, W.; Wu, X.Y.; Gao, M.X.; Bai, P.K. Effect of Selective Laser Melting Process Parameters on Microstructure and Properties of Co-Cr Alloy. *Materials* **2018**, *11*, 1546. [CrossRef]
2. Que, Z.Q.; Chang, L.T.; Saario, T.; Bojinov, M. Localised electrochemical processes on laser powder bed fused 316 stainless steel with various heat treatments in high-temperature water. *Addit. Manuf.* **2022**, *60*, 17. [CrossRef]

3. Cheng, B.; Shrestha, S.; Chou, K. Stress and deformation evaluations of scanning strategy effect in selective laser melting. *Addit. Manuf.* **2016**, *12*, 240–251. [CrossRef]
4. Kajima, Y.; Takaichi, A.; Kittikundecha, N.; Nakamoto, T.; Kimura, T.; Nomura, N.; Kawasaki, A.; Hanawa, T.; Takahashi, H.; Wakabayashi, N. Effect of heat-treatment temperature on microstructures and mechanical properties of Co–Cr–Mo alloys fabricated by selective laser melting. *Mater. Sci. Eng. A* **2018**, *726*, 21–31. [CrossRef]
5. Vandenbroucke, B.; Kruth, J.P. Selective laser melting of biocompatible metals for rapid manufacturing of medical parts. *Rapid Prototyp. J.* **2007**, *13*, 196–203. [CrossRef]
6. Koutsoukis, T.; Zinelis, S.; Eliades, G.; Al-Wazzan, K.; Al Rifaiy, M.; Al Jabbari, Y.S. Selective Laser Melting Technique of Co-Cr Dental Alloys: A Review of Structure and Properties and Comparative Analysis with Other Available Techniques. *J. Prosthodont.* **2015**, *24*, 303–312. [CrossRef] [PubMed]
7. Maconachie, T.; Leary, M.; Lozanovski, B.; Zhang, X.; Qian, M.; Faruque, O.; Brandt, M. SLM lattice structures: Properties, performance, applications and challenges. *Mater. Des.* **2019**, *183*, 108137. [CrossRef]
8. Taniguchi, N.; Fujibayashi, S.; Takemoto, M.; Sasaki, K.; Otsuki, B.; Nakamura, T.; Matsushita, T.; Kokubo, T.; Matsuda, S. Effect of pore size on bone ingrowth into porous titanium implants fabricated by additive manufacturing: An in vivo experiment. *Mater. Sci. Eng. C* **2016**, *59*, 690–701. [CrossRef] [PubMed]
9. Kluczyński, J.; Śnieżek, L.; Grzelak, K.; Mierzyński, J. The Influence of Exposure Energy Density on Porosity and Microhardness of the SLM Additive Manufactured Elements. *Materials* **2018**, *11*, 2304. [CrossRef]
10. Bartolomeu, F.; Costa, M.M.; Alves, N.; Miranda, G.; Silva, F.S. Selective Laser Melting of Ti6Al4V sub-millimetric cellular structures: Prediction of dimensional deviations and mechanical performance. *J. Mech. Behav. Biomed. Mater.* **2021**, *113*, 104123. [CrossRef]
11. Buj-Corral, I.; Tejo-Otero, A.; Fenollosa-Artés, F. Development of AM Technologies for Metals in the Sector of Medical Implants. *Metals* **2020**, *10*, 686. [CrossRef]
12. Balagna, C.; Spriano, S.; Faga, M.G. Characterization of Co-Cr-Mo alloys after a thermal treatment for high wear resistance. *Mater. Sci. Eng. C-Mater. Biol. Appl.* **2012**, *32*, 1868–1877. [CrossRef] [PubMed]
13. Wang, R.X.; Qin, G.W.; Zhang, E.L. Effect of Cu on Martensite Transformation of CoCrMo alloy for biomedical application. *J. Mater. Sci. Technol.* **2020**, *52*, 127–135. [CrossRef]
14. Lu, Y.J.; Zhao, W.; Yang, C.G.; Liu, Y.J.; Xiang, H.L.; Yang, K.; Lin, J.X. Improving mechanical properties of selective laser melted Co29Cr9W3Cu alloy by eliminating mesh-like random high-angle grain boundary. *Mater. Sci. Eng. A-Struct. Mater. Prop. Microstruct. Process.* **2020**, *793*, 12. [CrossRef]
15. Saini, J.S.; Dowling, L.; Trimble, D.; Singh, D. Mechanical Properties of Selective Laser Melted CoCr Alloys: A Review. *J. Mater. Eng. Perform.* **2021**, *30*, 8700–8714. [CrossRef]
16. Ammarullah, M.I.; Santoso, G.; Sugiharto, S.; Supriyono, T.; Kurdi, O.; Tauviqirrahman, M.; Winarni, T.; Jamari, J. Tresca stress study of CoCrMo-on-CoCrMo bearings based on body mass index using 2D computational model. *J. Tribol.* **2022**, *33*, 31–38.
17. Liu, Y.; Zhu, Q.; Wang, C.; Li, J. Tribological behavior of CoCrMo artificial knee joint with symmetrically biomimetic textured surfaces on PEEK. *Opt. Laser Technol.* **2023**, *157*, 108774. [CrossRef]
18. Murr, L.E.; Gaytan, S.M.; Martinez, E.; Medina, F.; Wicker, R.B. Next Generation Orthopaedic Implants by Additive Manufacturing Using Electron Beam Melting. *Int. J. Biomater.* **2012**, *2012*, 245727. [CrossRef]
19. Li, H.Q.; Wang, M.; Lou, D.J.; Xia, W.L.; Fang, X.Y. Microstructural features of biomedical cobalt-chromium-molybdenum (CoCrMo) alloy from powder bed fusion to aging heat treatment. *J. Mater. Sci. Technol.* **2020**, *45*, 146–156. [CrossRef]
20. Balagna, C.; Faga, M.G.; Spriano, S. Tribological behavior of a Ta-based coating on a Co-Cr-Mo alloy. *Surf. Coat. Technol.* **2014**, *258*, 1159–1170. [CrossRef]
21. Isik, M.; Avila, J.D.; Bandyopadhyay, A. Alumina and tricalcium phosphate added CoCr alloy for load-bearing implants. *Addit. Manuf.* **2020**, *36*, 8. [CrossRef] [PubMed]
22. Liu, S.F.; Li, Y.Z.; Wang, Y.; Wei, Y.K.; Zhang, L.L.; Wang, J.Y.; Yang, X. Selective laser melting of WC-Co reinforced AISI 1045 steel composites: Microstructure characterization and mechanical properties. *J. Mater. Res. Technol-JMRT* **2022**, *19*, 1821–1835. [CrossRef]
23. Lei, Y.; Bian, H.; Fu, W.; Song, X.; Feng, J.; Long, W.; Niu, H. Evaluation of Biomedical Ti/ZrO2 Joint Brazed with Pure Au Filler: Microstructure and Mechanical Properties. *Metals* **2020**, *10*, 526. [CrossRef]
24. Bian, H.; Song, X.; Hu, S.; Lei, Y.; Jiao, Y.; Duan, S.; Feng, J.; Long, W. Microstructure Evolution and Mechanical Properties of Titanium/Alumina Brazed Joints for Medical Implants. *Metals* **2019**, *9*, 644. [CrossRef]
25. Ammarullah, M.I.; Santoso, G.; Sugiharto, S.; Supriyono, T.; Wibowo, D.B.; Kurdi, O.; Tauviqirrahman, M.; Jamari, J. Minimizing Risk of Failure from Ceramic-on-Ceramic Total Hip Prosthesis by Selecting Ceramic Materials Based on Tresca Stress. *Sustainability* **2022**, *14*, 13413. [CrossRef]
26. Canpolat, O.; Canakci, A.; Erdemir, F. Evaluation of microstructure, mechanical, and corrosion properties of SS316L/Al2O3 composites produced by hot pressing. *Mater. Chem. Phys.* **2022**, *280*, 10. [CrossRef]
27. Jiang, J.; Xiao, G.; Che, C.; Wang, Y. Microstructure, Mechanical Properties and Wear Behavior of the Rheoformed 2024 Aluminum Matrix Composite Component Reinforced by Al_2O_3 Nanoparticles. *Metals* **2018**, *8*, 460. [CrossRef]
28. Dosta, S.; Cinca, N.; Silvello, A.; Cano, I.G. Alumina Reinforcement of Inconel 625 Coatings by Cold Gas Spraying. *Metals* **2020**, *10*, 1263. [CrossRef]

29. Mercado-Lemus, V.H.; Gomez-Esparza, C.D.; Díaz-Guillén, J.C.; Mayén-Chaires, J.; Gallegos-Melgar, A.; Arcos-Gutierrez, H.; Hernández-Hernández, M.; Garduño, I.E.; Betancourt-Cantera, J.A.; Perez-Bustamante, R. Wear Dry Behavior of the Al-6061-Al2O3 Composite Synthesized by Mechanical Alloying. *Metals* **2021**, *11*, 1652. [CrossRef]
30. Kumar, N.M.S.; Shashank, T.N.; Dhruthi. Review-Different Ceramic Reinforcements In Aluminium Metal Matrix Composites. *ECS J. Solid State Sci. Technol.* **2021**, *10*, 8. [CrossRef]
31. Jue, J.B.; Gu, D.D.; Chang, K.; Dai, D.H. Microstructure evolution and mechanical properties of Al-Al_2O_3 composites fabricated by selective laser melting. *Powder Technol.* **2017**, *310*, 80–91. [CrossRef]
32. Liao, H.; Zhu, J.; Chang, S.; Xue, G.; Zhu, H.; Chen, B. Al_2O_3 loss prediction model of selective laser melting Al_2O_3–Al composite. *Ceram. Int.* **2020**, *46*, 13414–13423. [CrossRef]
33. Li, M.; Fang, A.; Martinez-Franco, E.; Alvarado-Orozco, J.M.; Pei, Z.J.; Ma, C. Selective laser melting of metal matrix composites: Feedstock powder preparation by electroless plating. *Mater. Lett.* **2019**, *247*, 115–118. [CrossRef]
34. Cao, Y.; Liu, W.; Zhang, L.; Sun, H.; Bai, P. Preparation and properties of CoCrMo-Ni@Al_2O_3 composite by selective laser melting. *Mater. Charact.* **2022**, *194*, 112435. [CrossRef]
35. Parra, C.; Montero-Silva, F.; Gentil, D.; del Campo, V.; da Cunha, T.H.R.; Henriquez, R.; Haberle, P.; Garin, C.; Ramirez, C.; Fuentes, R.; et al. The Many Faces of Graphene as Protection Barrier. Performance under Microbial Corrosion and Ni Allergy Conditions. *Materials* **2017**, *10*, 1406. [CrossRef] [PubMed]
36. Zhu, L.; Guo, H.B.; Gong, S.K. Sub-micron Co-Al_2O_3 composite powders prepared by room-temperature ultrasonic-assisted electroless plating. *Rare Met.* **2022**, *41*, 1968–1971. [CrossRef]
37. Gao, J.B.; Jin, Y.T.; Fan, Y.Q.; Xu, D.K.; Meng, L.; Wang, C.; Yu, Y.P.; Zhang, D.L.; Wang, F.H. Fabricating antibacterial CoCrCuFeNi high-entropy alloy via selective laser melting and in-situ alloying. *J. Mater. Sci. Technol.* **2022**, *102*, 159–165. [CrossRef]
38. Gui, Z.; Zhu, J.; Hu, Y. A simple preparative method for porous hollow-ware stacked cobalt oxides and their catalytic properties. *Mater. Chem. Phys.* **2010**, *124*, 243–247. [CrossRef]
39. Liu, X.H.; Yi, R.; Zhang, N.; Shi, R.R.; Li, X.G.; Qiu, G.Z. Cobalt hydroxide nanosheets and their thermal decomposition to cobalt oxide nanorings. *Chem.-Asian J.* **2008**, *3*, 732–738. [CrossRef]
40. Kajima, Y.; Takaichi, A.; Htata, H.L.; Hanawa, T.; Wakabayashi, N. Recrystallization behavior of selective laser melted Co-Cr-Mo alloys with several heat treatment times. *Mater. Sci. Eng. A-Struct. Mater. Prop. Microstruct. Process.* **2022**, *856*, 10. [CrossRef]
41. Li, K.F.; Wang, Z.; Song, K.K.; Khanlari, K.; Yang, X.S.; Shi, Q.; Liu, X.; Mao, X.H. Additive manufacturing of a Co-Cr-W alloy by selective laser melting: In-situ oxidation, precipitation and the corresponding strengthening effects. *J. Mater. Sci. Technol.* **2022**, *125*, 171–181. [CrossRef]
42. Yin, J.N.; Liu, W.; Cao, Y.; Zhang, L.H.; Wang, J.H.; Li, Z.H.; Zhao, Z.Y.; Bai, P.K. Rapid prediction of the relationship between processing parameters and molten pool during selective laser melting of cobalt-chromium alloy powder: Simulation and experiment. *J. Alloy. Compd.* **2022**, *892*, 11. [CrossRef]
43. Abolhasani, D.; Seyedkashi, S.M.H.; Kang, N.; Kim, Y.J.; Woo, Y.Y.; Moon, Y.H. Analysis of Melt-Pool Behaviors during Selective Laser Melting of AISI 304 Stainless-Steel Composites. *Metals* **2019**, *9*, 876. [CrossRef]
44. Zhang, J.; Song, B.; Cai, C.; Zhang, L.; Shi, Y. Tailorable microstructure and mechanical properties of selective laser melted TiB/Ti–6Al–4V composite by heat treatment. *Adv. Powder Mater.* **2022**, *1*, 100010. [CrossRef]
45. Li, X.W.; Willy, H.J.; Chang, S.; Lu, W.H.; Herng, T.S.; Ding, J. Selective laser melting of stainless steel and alumina composite: Experimental and simulation studies on processing parameters, microstructure and mechanical properties. *Mater. Des.* **2018**, *145*, 1–10. [CrossRef]
46. Takaichi, A.; Suyalatu; Nakamoto, T.; Joko, N.; Nomura, N.; Tsutsumi, Y.; Migita, S.; Doi, H.; Kurosu, S.; Chiba, A.; et al. Microstructures and mechanical properties of Co-29Cr-6Mo alloy fabricated by selective laser melting process for dental applications. *J. Mech. Behav. Biomed. Mater.* **2013**, *21*, 67–76. [CrossRef]
47. Dong, X.; Li, N.; Zhou, Y.N.; Peng, H.B.; Qu, Y.T.; Sun, Q.; Shi, H.J.; Li, R.; Xu, S.; Yan, J.Z. Grain boundary character and stress corrosion cracking behavior of Co-Cr alloy fabricated by selective laser melting. *J. Mater. Sci. Technol.* **2021**, *93*, 244–253. [CrossRef]
48. Cui, G.J.; Liu, Y.P.; Li, S.; Liu, H.Q.; Gao, G.J.; Kou, Z.M. Nano-TiO_2 reinforced CoCr matrix wear resistant composites and high-temperature tribological behaviors under unlubricated condition. *Sci. Rep.* **2020**, *10*, 12. [CrossRef]
49. Hong, Q.; Bai, P.; Wang, J. Effects of Al_2O_3 Addition on the Microstructure and Properties of CoCr Alloys. *Metals* **2019**, *9*, 1074. [CrossRef]
50. Kale, A.B.; Singh, J.; Kim, B.K.; Kim, D.I.; Choi, S.H. Effect of initial microstructure on the deformation heterogeneities of 316L stainless steels fabricated by selective laser melting processing. *J. Mater. Res. Technol.-JMRT* **2020**, *9*, 8867–8883. [CrossRef]

metals

MDPI

Article

Microstructure and High Temperature-Mechanical Properties of TiC/Graphene/Ti6Al4V Composite Formed by Laser Powder Bed Fusion

Shijie Chang [1], Wenbo Du [2,*], Zhanyong Zhao [1,*] and Peikang Bai [1]

[1] School of Materials Science and Engineering, North University of China, Taiyuan 030051, China
[2] National Key Laboratory for Remanufacturing, Academy of Army Armored Forces, Beijing 100072, China
* Correspondence: dwbneu@163.com (W.D.); zhaozy@nuc.edu.cn (Z.Z.); Tel.: +86-136-2346-6690 (Z.Z.)

Abstract: TiC/graphene/Ti6Al4V composites were prepared by laser powder bed fusion using graphene and Ti6Al4V powder. The differences in microstructure and high-temperature mechanical properties between the Ti6Al4V alloy and the TiC/graphene/Ti6Al4V composite were studied. The tensile and microhardness of the two materials were tested at 400 °C, 500 °C, and 600 °C; the results of the TiC/graphene/Ti6Al4V composite were 126 MPa, 162 MPa, and 76 MPa and 70 HV, 59 HV, and 61HV, respectively, higher than those of the Ti6Al4V alloy. These results happened because graphene reacted with Ti to form TiC particles, which were homogeneously distributed amongst α' acicular martensite. The addition of graphene refined the size of the acicular α' martensite. At the same time, the graphene and TiC particles showed a dispersion-strengthening effect. The mechanical properties of the TiC/graphene/Ti6Al4V composite were improved by the combination of fine-grain strengthening and dispersion strengthening mechanisms.

Keywords: laser powder bed fusion; graphene; high temperature; mechanical property

Citation: Chang, S.; Du, W.; Zhao, Z.; Bai, P. Microstructure and High Temperature-Mechanical Properties of TiC/Graphene/Ti6Al4V Composite Formed by Laser Powder Bed Fusion. *Metals* **2023**, *13*, 163. https://doi.org/10.3390/met13010163

Academic Editors: Milan Brandt and Aleksander Lisiecki

Received: 20 October 2022
Revised: 8 January 2023
Accepted: 11 January 2023
Published: 12 January 2023

1. Introduction

Titanium alloy has the advantages of high specific and thermal strength, strong corrosion resistance, and good process performance, and it has become an ideal engineering structural material in aerospace, chemical, and other fields [1]. The Ti6Al4V alloy can work at a temperature of 400 °C for a long time and is mainly used to manufacture fans, compressor disks, and engine blades in the aviation industry [2]. However, with the development of the new generation of aerospace technology, the high-temperature properties of the Ti6Al4V alloy no longer meet the requirements, whereas titanium matrix composites have become candidate materials for applications in high-speed aerospace vehicles and new aeroengines [3,4]. Because ceramic particles have high hardness, high wear resistance, and high-temperature performance, they are widely used to manufacture titanium matrix composites, such as TiC [5], WC [6], and TiB [7]. In recent years, because of its unique electrical, mechanical, and thermal properties, graphene has become an ideal reinforcement material for preparing lightweight, high-strength, and high-performance metal matrix composites [8]. For example, Guo et al. [9] developed the Ni-P@GNFs/Ti6Al4V composite. The compressive strength of the composite reached 1133 MPa, an increase of 30.6% compared to that of Ti6Al4V (867 MPa). Shang et al. [10] also used plasma sintering to prepare multi-layer graphene and Ti6Al4V-based discontinuous reinforced composite that reached a tensile strength of 1010 MPa [10].

Laser powder bed fusion (LPBF) is a powder bed melting additive manufacturing (AM) technology that forms parts by layer-by-layer selective melting and layer-by-layer superposition [10,11]. Based on the forming characteristics, LPBF technology can integrate parts with complex structures, shorten the product development cycle, and improve the utilization of materials [12,13]. In recent years, LPBF technology has been widely used in medical, dental, and especially aerospace fields [14], and the parts formed with porous structures have reduced weight [15]. Although LPBF technology has many advantages, there are still some defects such as pores and residual stress; hence, the quality of products can be improved by adjusting the printing parameters to control such defects [16–18], especially by combining large data acquisition and analysis with the printing process [19]. With the use of LPBF technology, graphene-reinforced metal matrix materials can significantly improve the performance of composites. Zhao et al. [20] used an organic aluminum reduction method to coat graphene with a nano aluminum layer and used LPBF technology to prepare aluminum-coated graphene/AlSi10Mg alloy matrix composites and the nano-scale Al4C3 precipitates generated in the composite. The tensile strength and hardness of the composite increased by 11% and 40.8%, respectively. Wang et al. [21] used the LPBF technology to prepare a graphene/Inconel 718 composite, and the tensile strength and Young's modulus of the composite reached 1511 MPa and 675 GPa, respectively [21].

Previous studies have used LPBF technology to enhance Ti6Al4V alloys with graphene, which exhibits excellent mechanical properties at room temperature [22–24]. However, the high-temperature mechanical properties of TiC/Ti6Al4V composites have rarely been reported. In this study, the TiC/graphene/Ti6Al4V composites were prepared by the LPBF method using Ti6Al4V alloy powder and graphene powder as raw materials. The purpose of this study was to explore the differences between the microstructure and high-temperature mechanical properties of the LPBF-formed Ti6Al4V alloy and the TiC/graphene/Ti6Al4V composite to reveal the strengthening mechanism of the composite.

2. Materials and Methods

The raw material for preparing the composite materials was Ti6Al4V alloy powder provided by Zhonghang Maite Powder Metallurgy Technology Co. Ltd. (Beijing, China). The particle size distribution of the powder was D10 = 19.9 μm, D50 = 35.4 μm, D90 = 58 μm. The powder material composition, provided by the company, is shown in Table 1. Note that 0.3-wt% graphene was added to the composite materials. The graphene powder was provided by KENA Carbon New Materials Co. Ltd. (Lanzhou, China), and the graphene morphology is shown in Figure 1a.

Table 1. Chemical composition of Ti6Al4V Powder.

Ti	Al	V	Fe	C	N	H	O	Si
bal	5.5~6.5	3.5~4.5	≤0.25	≤0.08	≤0.03	≤0.012	≤0.1	<0.15

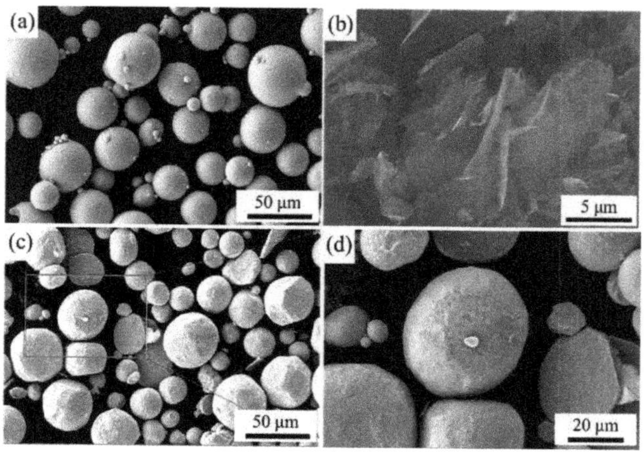

Figure 1. (**a**) Ti6Al4V powder morphology; (**b**) graphene morphology; (**c**,**d**) morphology of graphene and Ti6Al4V powder after ball milling.

A planetary ball mill (QM-3SP4 model) was adopted. The ball milling parameters were as follows: the specific gravity between the weight of the ball and the weight of the composite powder was 8:1; the ball milling time was 3 h; and the ball milling speed was 240 rad/min. The mixed powder is shown in Figure 1 b. The metal 3D printing equipment of Easy 3D EP-M150 was used, and the two materials used the same printing process parameters. The process parameters used were as follows: laser power of 200 W, scanning spacing of 0.12 mm, powder layer thickness of 0.03 mm, and scanning speed of 1200 mm/s; a checkerboard scanning strategy was adopted, and each layer was rotated by 67°. The substrate was preheated to 120 °C. The sample forming building direction is shown in Figure 2a.

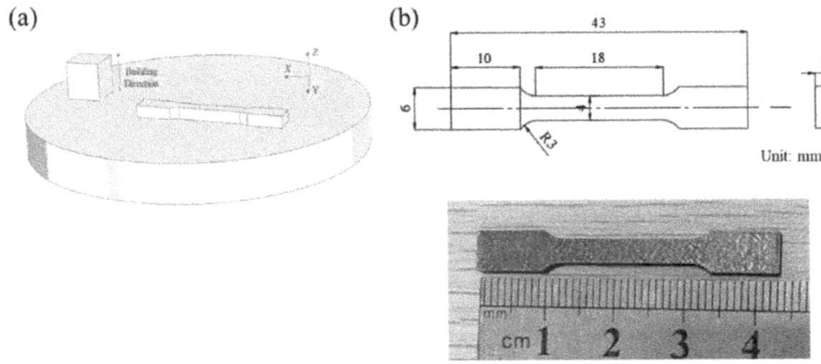

Figure 2. (**a**) Schematic of building direction; (**b**) tensile specimen.

An MH-300A electronic densitometer was used to measure the density of the samples. Five samples of each material were measured, and the average value ρs was taken. The density value was calculated as ρ_s/ρ_n * 100. The ρ_n is the standard density of Ti-x vol.% graphene, and the density of Ti6Al4V is 4.5g/cm^3. The HGW-300G microcomputer-controlled high-temperature material tensile testing machine was used to raise the temperature of the sample to 400 °C, 500 °C, and 600 °C at a rate of 10 °C/s and to keep it stable for 5 min. The tensile rate was set as 0.5 mm/min. The experiment was repeated three times and the average value was used. The size of the tensile sample is shown in Figure 2b. Thereafter, the ZD-HVZHT-30 high-temperature Vickers hardness tester was used to raise the temperature of the sample to 400 °C, 500 °C, and 600 °C at a heating rate of 10 °C/s, and to keep it stable for 5 min. A diamond indenter with a load of 4.904 N (500 g) was used for 10 s. Five points were randomly selected on the polished sample surface for the measurements, and the average values of these five measurements were considered.

Next, the standard polished parts were placed in a 5 mL HF, 25 mL HNO3, and 50 mL H2O corrosion solution for etching for 30–35 s. A field emission scanning electron microscope (SEM; JSM-7900F, JEOL, Peabody, MA, USA) was used to observe the structure of the sample with a working voltage set to 10.0 KV. The preparation of the TEM sample was as follows: First, the sample was cut into 0.5 mm thick sheets with a wire cutting machine, and the sheets were ground to 35~40 μm with 1000 # SiC water abrasive paper thick. Then, the sheet was punched into a round shape with a diameter of 3 mm. Finally, a Gatan-691 precision argon ion instrument was used for thinning. TEM samples were characterized by field emission transmission electron microscopy (JEM-2100F) with an acceleration voltage of 200kV.

3. Results and Discussion

3.1. Initial Microstructure

Figure 3 shows the microstructure of the Ti6Al4V alloy and graphene-reinforced Ti6Al4V composite. A large amount of coarse α′ acicular martensite was observed. The α′ acicular martensite was refined, as shown in Figure 3. New nanoparticles were formed in the graphene-reinforced Ti6Al4V composite, as shown in Figure 3d. According to the EDS analysis (Figure 3f), the granular precipitation was mainly composed of Ti and C elements, and the atomic ratio was close to 1:1. Figure 3d shows the TEM micromorphology and selected area electron diffraction (SAED) (Figure 3e) of TiC particles. Previous studies have also verified the existence of graphene [25] and TiC (Figure 4 [26]). In general, the reinforcing phases in the graphene-reinforced Ti6Al4V composite are graphene and TiC particles.

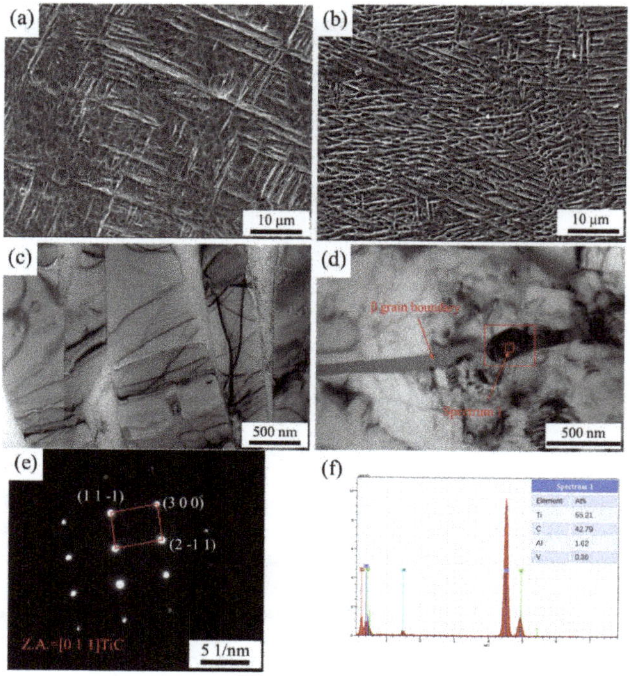

Figure 3. (**a**) SEM image of the longitudinal section of the Ti6Al4V alloy; (**b**) SEM images of the graphene reinforced Ti6Al4V composite; (**c**) TEM morphology of the Ti6Al4V alloy; (**d**) TEM morphology of the graphene reinforced Ti6Al4 V composite; (**e**) SAED figure of the selected area in (**d**); (**f**) composition analysis at point 1.

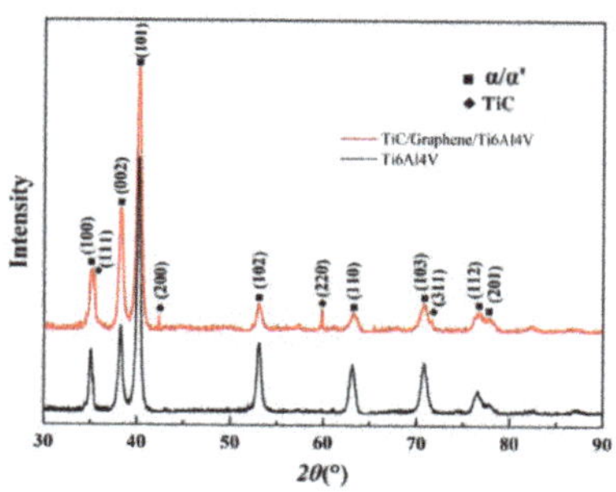

Figure 4. XRD patterns of the TiC/graphene/Ti6Al4V composite and Ti6Al4V alloy. Reprinted with permission from Ref. [26]. 2023, Elsevier.

Figure 5 shows the micro morphologies of the Ti6Al4V alloy and composite at a lower magnification under SEM, where no pores were observed, while a small number of pores with a diameter of 5 μ m were observed in the composite. The density of the

Ti6Al4V alloy was 99.2%, and that of the TiC/graphene/Ti6Al4V composite was 98.9%. Therefore, the porosity of the formed composite material was more than that of the Ti6Al4V alloy; however, a density above 98% had little influence on its mechanical properties [27]. Therefore, the influence of porosity on the formation of voids during tension is lower.

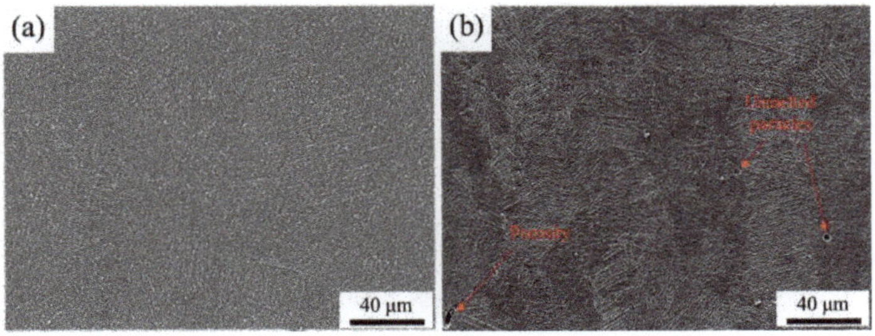

Figure 5. SEM microtopography; (**a**) Ti6Al4V; (**b**) TiC/graphene/Ti6Al4V.

3.2. High-Temperature Performance

Figure 6 shows the engineering tensile strength of the LPBF-formed Ti6Al4V and TiC/graphene/Ti6Al4V composite at 400 °C, 500 °C, and 600 °C. At 400 °C, the tensile strengths of the Ti6Al4V alloy and composite were 880 MPa and 1006 MPa, respectively, which was 126 MPa higher than that of the Ti6Al4V alloy. At 500 °C, the tensile strengths of the Ti6Al4V alloy and composite were 806 MPa and 162 MPa higher than that of the Ti6Al4V alloy, respectively. At 600 °C, the tensile strengths of the Ti6Al4V alloy and composite were 462 MPa and 76 MPa higher than that of the Ti6Al4V alloy, respectively.

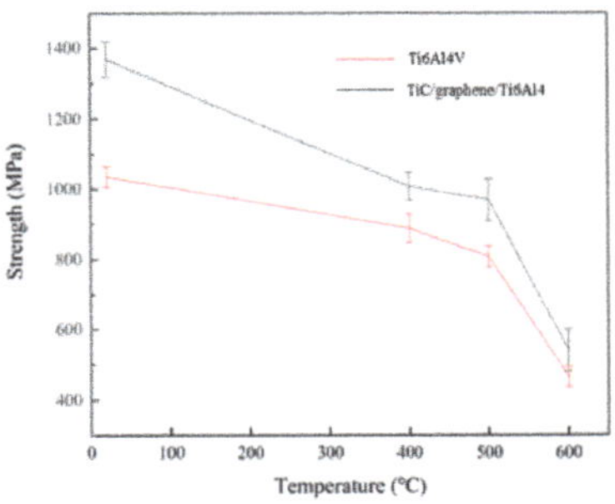

Figure 6. Tensile strength at different temperatures.

Figure 7 shows the fractographic morphology at the same temperature for both materials, with an increase in the dissociation surface area of the TiC/graphene/Ti6Al4V composites at a fractographic morphology of 400 °C and the formation of fewer pores during the drawing of the Ti6Al4V alloy. The area of the dissociation surface of both materials fractioned at 500 °C decreased compared to that at 400 °C, but the number of pores with a fracture morphology at 500 °C increased in the composite compared to that in the Ti6Al4V alloy, while the size of the pores decreased. The fracture appearances of the two materials differed greatly at 600 °C, and the pore size and number of Ti6Al4V alloy were larger than that of the composite material because the softening behavior of the Ti6Al4V alloy at 600 °C was increased as well as the number of holes in the fracture appearances of the alloy. However, the composites contained reinforcement relative to the material; thus, the size and number of holes in the composite were smaller than those in the Ti6Al4V alloy, and the reinforcement phase effect was also reflected in the fracture morphology at 400 °C and 500 °C.

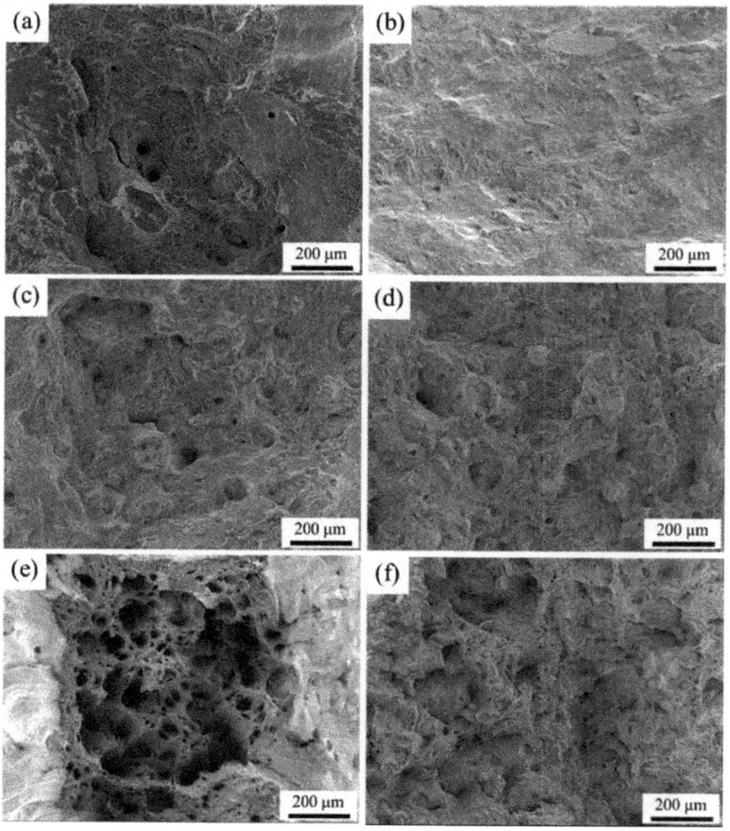

Figure 7. Fracture morphologies at different temperatures: (**a**,**c**,**e**) fracture morphologies of Ti6Al4V alloys at 400 °C, 500 °C, and 600 °C; (**b**,**d**,**f**) fracture morphologies of TiC/graphene/Ti6Al4V composite at 400 °C, 500 °C, and 600 °C.

Further analysis of the fracture mechanism of materials at high temperatures was carried out using micromorphology. The tensile fracture morphology of the Ti6Al4V alloy at 400 °C is shown in Figure 8a, with an average size of statistical dimple of 4.5 μm. The dimples, tear edges on the outside of the dimples, and a large step-like dissociation surface indicate that the Ti6Al4V alloy underwent both a ductile fracture and a brittle fracture at 400 °C. The dimple size of the Ti6Al4V alloy in the fracture morphology at 500 °C was counted, and the average dimple size was 4.2 μm; the dimple characteristics are similar to those of the fracture at 400 °C. The fracture morphology of the high-resolution SEM at 600 °C is shown in Figure 8c,d. The morphology inside the hole is also a dimple, with an average statistical dimple size of 0.6 μm.

The reason that the fracture morphology at 400 °C contained a large dissociation surface was that the Ti6Al4V alloy is mainly composed of α phase, which has a hexagonal structure, there are fewer slip systems in the phase, and the energy required for the dislocation slip is higher. In the fracture morphology at 500 °C, the reason for the reduction of the dissociation surface is mainly that the temperature increase provided more energy for the slip system to start easier; therefore, the dissociation surface was reduced. The formation of dimples is mainly due to the growth, aggregation, and fracture of micro-voids generated inside the alloy during the deformation. The micro-cavities of most alloys are formed at the second phase particles, while the Ti6Al4V alloy is a two-phase alloy, and the content of the phase structure is very small; therefore, the micro-cavities of the Ti6Al4V alloy were mainly formed at the grain boundary [28]. With the increase in the tensile temperature, the proportion of dimples in the fracture morphology increased significantly. The reason for this is also that the energy of the externally supplied material increases with increasing temperature, making the dislocation more prone to slip. Slipping at a large number of dislocations accelerates the growth, aggregation, and fracture of the micro-voids, resulting in a larger void size as shown in Figure 7, reducing the size of the dimples, and decreasing the tensile strength of the Ti6Al4V alloy with increasing temperature.

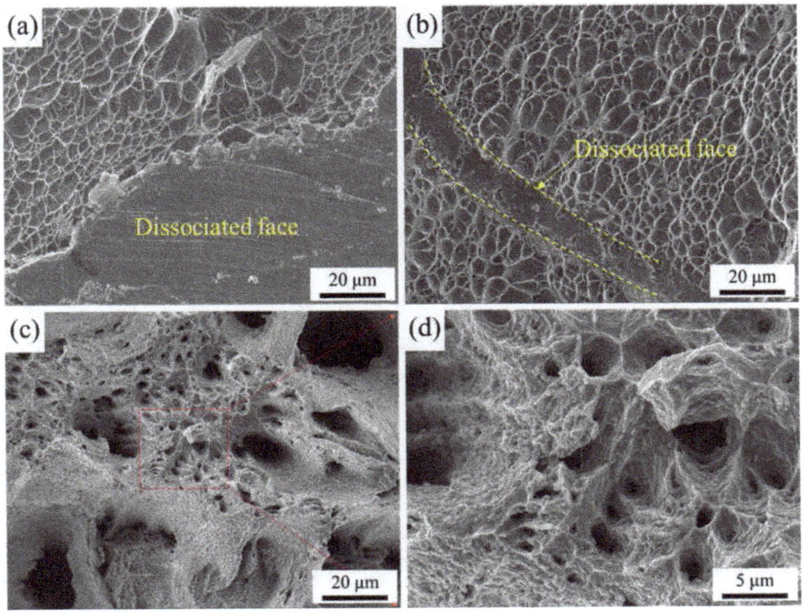

Figure 8. The high-temperature tensile fracture image of the Ti6Al4V alloy: (**a**) Fracture morphology at 400 °C; (**b**) fracture morphology at 500 °C; (**c**,**d**) Fracture morphology at 600 °C.

The fracture shape of the TiC/graphene/Ti6Al4V composite at 400 °C is shown in Figure 9a. Large cracks and a large number of dissociation surfaces are visible. The large cracks revealed a 'u' shape; at the same time, the sheet structure in the hole shown in Figure 9b–e, enlarged sheet structure of Figure 9e, and EDS surface sweep elemental analysis showed that the lamellar structure consisted of carbon elements as shown in Figure 9f. Therefore, the lamellar structure is unreacted graphene. The fracture morphology on both sides of the hole was different: one side was the dissociation surface, and the other side was the dimple (the average size of statistical dimples is 1.1 μm), as shown in Figure 9b. The fracture morphology of the TiC/graphene/Ti6Al4V composite at 500 °C shows a large number of dimples (average size of 0.97 μm) and a small amount of dissociation surface, while the dimple size of the TiC/graphene/Ti6Al4V composite is significantly reduced, and the dissociation surface is smaller than that of the Ti6Al4V alloy. The same sheet structure as that shown in Figure 9 existed in the holes, as shown in Figure 10c, and the composite of the sheet structure was carbon, as confirmed by an EDS surface scanning analysis. Therefore, this sheet structure was unreacted graphene in the composite. The fracture morphology of the TiC/graphene/Ti6Al4V composite at 600 °C was also the same as that of the Ti6Al4V alloy in Figure 10f, and the number and size of the pores were much smaller than those of the Ti6Al4V alloy, evidenced by the statistical dimple size average of 0.54 μm.

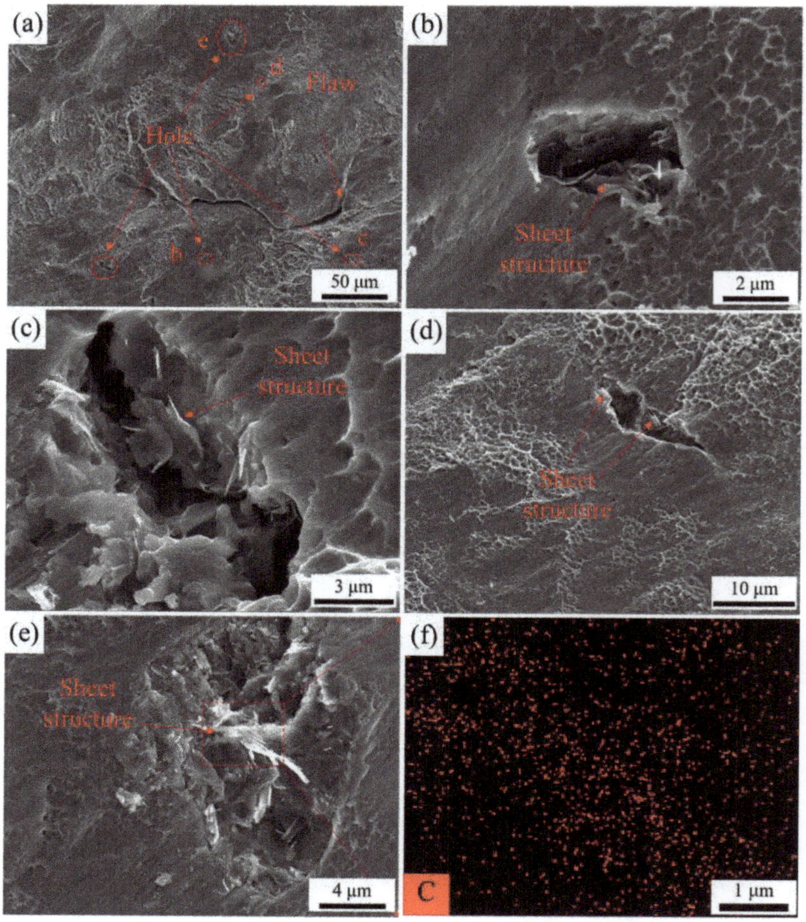

Figure 9. High-temperature tensile fracture of the TiC/graphene/Ti6Al4V composite: (**a**) 400 °C fracture; (**b–f**) enlarged morphology of b–f in (**a**); (**f**) distribution of element C in the selected area of ©.

By comparing the average dimple size of the Ti6Al4V alloy mentioned above, the dimple size of the TiC/graphene/Ti6Al4V composite material is significantly smaller at the same temperature. The decrease in dimple size is related to the grain refinement, which improves the performance of the material and precipitation of the composite material. The size of fracture dimples of both materials decreased gradually from 400–600 °C because the dislocation slip accelerates with the increase in temperature, and the growth, aggregation, and fracture speed of micro-voids increases, which leads to the decrease in dimple size and performance with the increase in temperature.

Graphene can be retained during LPBF by repeated rapid material melting and curing cycles. In addition, when a dislocation slip occurs, graphene prevents the dislocation slip and enhances the properties of the composite [29]. It was inferred from the above that the micro-cavities of the Ti6Al4V alloy were easy to form at the grain boundary and second phase particles. Therefore, the formation of voids with graphene also proves that, during the deformation, the micro-voids are easy to form together by precipitation and break with the growth and aggregation of micro-voids, as shown in Figures 9 and 10. The TiC in the TiC/graphene/Ti6Al4V composites was produced by an in situ reaction of graphene and Ti, as shown in Figure 3. The TiC particles enhanced the properties of the composite through the Orowan strengthening mechanism [30]. At the same time, nano-scale TiC particles were distributed in the composite matrix, which played a role in passivating the crack tip, delaying the crack growth [31], and enhancing the tensile properties of the composite. Thus, both the TiC particles and the graphene in the TiC/graphene/Ti6Al4V composite improved the tensile strength of the composite, making the tensile strength of the composite higher than that of the Ti6Al4V alloy at 400 °C, 500 °C, and 600 °C.

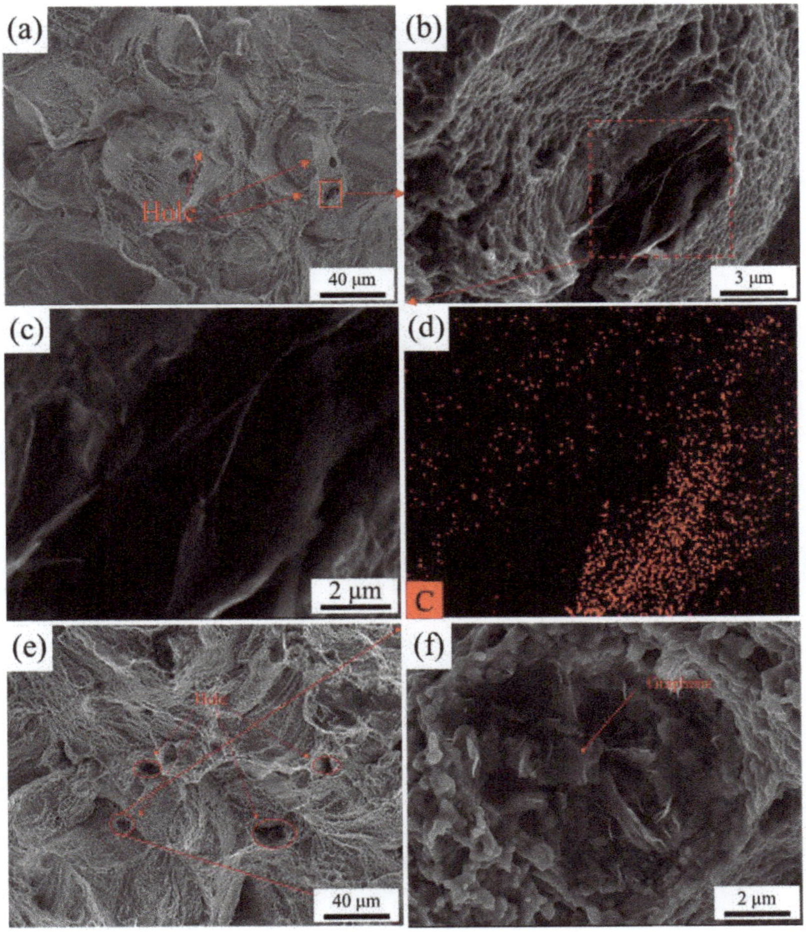

Figure 10. High-temperature tensile fracture of the TiC/graphene/Ti6Al4V composite: (**a**) fracture at 500 °C; (**b**) enlarged view of Figure (**a**); (**c**) enlarged view of (**b**); (**d**) the distribution of C elements on EDS surface of (**c**); (**e**,**f**) fracture morphology at 600 °C.

The microhardness of the Ti6Al4V alloy and TiC/graphene/Ti6Al4V composite formed by LPBF at room temperature and high temperatures is shown in Figure 11. The hardness of the two materials decreased step by step at high temperatures. The microhardness of the Ti6Al4V alloy at room temperature was 336 HV, and that formed by electron beam was 340 HV [32], whereas the microhardness of the TiC/graphene/Ti6Al4V composite was 419 HV, which was 83 HV higher than the LPBF-formed Ti6Al4V alloy and 79 HV higher than the Ti6Al4V alloy formed by the electron beam. The micro-hardness of the TiC/graphene/Ti6Al4V composite at 400 °C, 500 °C, and 600 °C was 70 HV,

59 HV, and 61 HV higher than that of the TiC/graphene/Ti6Al4V composite, respectively. The microhardness of the TiC/graphene/Ti6Al4V composite at 600 °C was higher than that of the Ti6Al4V alloy at 400 °C, indicating that the working temperature of the TiC/graphene/Ti6Al4V composite could be higher than that of the Ti6Al4V alloy. The hardness of the TiC/graphene/Ti6Al4V composite could be improved by the grain size refinement of the composite and dispersion strengthening mechanism of the TiC particles and graphene.

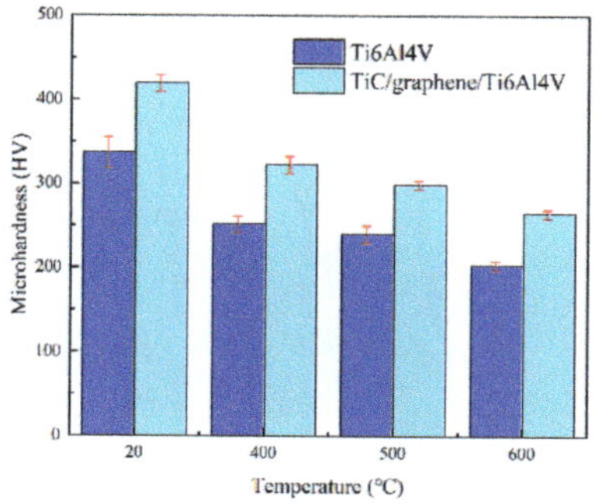

Figure 11. Microhardness of Ti6Al4V and TiC/graphene/Ti6Al4V alloys at room temperature and high temperatures.

4. Conclusions

In this study, the microstructure and high-temperature mechanical properties of Ti6Al4V and TiC/graphene/Ti6Al4V composites formed using LPBF were investigated and compared. The following conclusions were obtained:

(1) During the LPBF, graphene reacted with Ti in situ to form TiC particles, which were homogeneously distributed between α' acicular martensite; the addition of graphene refined the size of acicular α' martensite. At the same time, the graphene and the TiC particles played the role of dispersion strengthening, and the mechanical properties of the TiC/graphene/Ti6Al4V composite were improved by the combination of the fine-grain strengthening mechanism and the dispersion strengthening mechanism.

(2) The density of the Ti6Al4V alloy was 99.2% and that of the TiC/graphene/Ti6Al4V composite was 98.9%. However, a density above 98% has little influence on its mechanical properties [27].

(3) The high-temperature tensile strength of the TiC/graphene/Ti6Al4V composite sample was 126 MPa, 162 MPa, and 76 MPa higher than that of the Ti6Al4V material sample at 400 °C, 500 °C, and 600 °C, respectively.

(4) The addition of graphene refined the needle shape α' martensite size, while the average dimple size of the composite at the same temperature was smaller than that of the Ti6Al4V alloy. Graphene can be retained during LPBF by repeated rapid material melting and curing cycles, with microscopic voids formed during high-temperature tension.

(5) The microhardness of the TiC/graphene/Ti6Al4V composite at room temperature, 400 °C, 500 °C, and 600 °C was 82 HV, 70 HV, 59 HV, and 61 HV higher than that of the Ti6Al4V alloy, respectively.

Author Contributions: Conceptualization, Z.Z. and S.C.; methodology, S.C.; software, S.C.; validation, Z.Z., P.B. and W.D.; writing—original draft preparation, S.C.; writing—review and editing, Z.Z., P.B. and W.D.; supervision, Z.Z. All authors have read and agreed to the published version of the manuscript.

Funding: The authors would like to thank the National Natural Science Foundation of China (51775521, U22A20192), the Key Research and Development Project of Shanxi Province (201903D121009), Scientific and Technological Innovation Programs of Higher Education Institutions in Shanxi, Equipment Pre-research Field Fund, China (No.61400040208).

Data Availability Statement: The data presented in this study are presented in Figures and tables in this article.

Conflicts of Interest: The authors declare no conflict of interest.

References

1. Zhao, Z.; Zhang, L.; Bai, P.; Du, W.; Wang, S.; Xu, X.; Dong, Q.; Li, Y.; Han, B. Tribological Behavior of In Situ TiC/Graphene/Graphite/Ti6Al4V Matrix Composite Through Laser Cladding. *Acta Metall. Sin. Engl. Lett.* **2021**, *34*, 1317–1330. [CrossRef]
2. Christoph Leyens, M.P. *Titanium and Titanium Alloys: Fundamentals and Applications*; Chemical Industry Press: Beijing, China, 2003.
3. Zhou, Z.; Liu, Y.; Liu, X.; Zhan, Q.; Wang, K. Microstructure evolution and mechanical properties of in-situ Ti6Al4V–TiB composites manufactured by selective laser melting. *Compos. Part B Eng.* **2021**, *207*, 108567. [CrossRef]
4. Zhao, Z.; Bai, P.; Du, W.; Liu, B.; Pan, D.; Das, R.; Liu, C.; Guo, Z. An overview of graphene and its derivatives reinforced metal matrix composites: Preparation, properties and applications. *Carbon* **2020**, *170*, 302–326. [CrossRef]
5. Yu, C.; Liu, X.; Li, Y.; Song, C.; Ma, G.; Niu, F.; Wu, D. Investigations of the microstructure and performance of TiCp/Ti6Al4V composites prepared by directed laser deposition. *Int. J. Mech. Sci.* **2021**, *205*, 106595. [CrossRef]
6. Liang, X.; Liu, Z.; Wang, B. Physic-chemical analysis for high-temperature tribology of WC-6Co against Ti–6Al–4V by pin-on-disc method. *Tribol. Int.* **2020**, *146*, 106242. [CrossRef]
7. Zhang, W.; Wang, M.; Chen, W.; Feng, Y.; Yu, Y. Preparation of TiBw/Ti–6Al–4V composite with an inhomogeneous reinforced structure by a canned hot extrusion process. *J. Alloys Compd.* **2016**, *669*, 79–90. [CrossRef]
8. Yang, M.; Liu, Y.; Fan, T.; Zhang, D. Metal-graphene interfaces in epitaxial and bulk systems: A review. *Prog. Mater. Sci.* **2020**, *110*, 100652. [CrossRef]
9. Guo, Y.; Yu, K.; Niu, J.; Sun, M.; Dai, G.; Sun, Z.; Chang, H. Effect of reinforcement content on microstructures and mechanical properties of graphene nanoflakes-reinforced titanium alloy matrix composites. *J. Mater. Res. Technol.* **2021**, *15*, 6871–6882. [CrossRef]
10. Waqar, S.; Sun, Q.; Liu, J.; Guo, K.; Sun, J. Numerical investigation of thermal behavior and melt pool morphology in multi-track multi-layer selective laser melting of the 316L steel. *Int. J. Adv. Manuf. Technol.* **2020**, *112*, 879–895. [CrossRef]
11. Waqar, S.; Guo, K.; Sun, J. FEM analysis of thermal and residual stress profile in selective laser melting of 316L stainless steel. *J. Manuf. Process.* **2021**, *66*, 81–100. [CrossRef]
12. Ahmed Obeidi, M.; Uí Mhurchadha, S.M.; Raghavendra, R.; Conway, A.; Souto, C.; Tormey, D.; Ahad, I.U.; Brabazon, D. Comparison of the porosity and mechanical performance of 316L stainless steel manufactured on different laser powder bed fusion metal additive manufacturing machines. *J. Mater. Res. Technol.* **2021**, *13*, 2361–2374. [CrossRef]
13. Korkmaz, M.E.; Waqar, S.; Garcia-Collado, A.; Gupta, M.K.; Krolczyk, G.M. A technical overview of metallic parts in hybrid additive manufacturing industry. *J. Mater. Res. Technol.* **2022**, *18*, 384–395. [CrossRef]
14. Korkmaz, M.E.; Gupta, M.K.; Waqar, S.; Kuntoğlu, M.; Krolczyk, G.M.; Maruda, R.W.; Pimenov, D.Y. A short review on thermal treatments of Titanium & Nickel based alloys processed by selective laser melting. *J. Mater. Res. Technol.* **2022**, *16*, 1090–1101.
15. Korkmaz, M.E.; Gupta, M.K.; Robak, G.; Moj, K.; Krolczyk, G.M.; Kuntoğlu, M. Development of lattice structure with selective laser melting process: A state of the art on properties, future trends and challenges. *J. Manuf. Process.* **2022**, *81*, 1040–1063. [CrossRef]
16. Grasso, M.; Colosimo, B.M. Process defects andin situmonitoring methods in metal powder bed fusion: A review. *Meas. Sci. Technol.* **2017**, *28*, 044005. [CrossRef]
17. Vilaro, T.; Colin, C.; Bartout, J.D. As-Fabricated and Heat-Treated Microstructures of the Ti-6Al-4V Alloy Processed by Selective Laser Melting. *Metall. Mater. Trans. A* **2011**, *42*, 3190–3199. [CrossRef]
18. Panwisawas, C.; Qiu, C.L.; Sovani, Y.; Brooks, J.W.; Attallah, M.M.; Basoalto, H.C. On the role of thermal fluid dynamics into the evolution of porosity during selective laser melting. *Scr. Mater.* **2015**, *105*, 14–17. [CrossRef]
19. Majeed, A.; Zhang, Y.; Ren, S.; Lv, J.; Peng, T.; Waqar, S.; Yin, E. A big data-driven framework for sustainable and smart additive manufacturing. *Robot. Comput. Integr. Manuf.* **2021**, *67*, 102026. [CrossRef]
20. Zhao, Z.; Bai, P.; Misra, R.D.K.; Dong, M.; Guan, R.; Li, Y.; Zhang, J.; Tan, L.; Gao, J.; Ding, T.; et al. AlSi10Mg alloy nanocomposites reinforced with aluminum-coated graphene: Selective laser melting, interfacial microstructure and property analysis. *J. Alloys Compd.* **2019**, *792*, 203–214. [CrossRef]
21. Wang, Y.C.; Shi, J.; Lu, S.Q.; Wang, Y. Selective Laser Melting of Graphene-Reinforced Inconel 718 Superalloy: Evaluation of Microstructure and Tensile Performance. *J. Manuf. Sci. Eng. Trans. Asme* **2017**, *139*, 4. [CrossRef]
22. Huo, P.; Zhao, Z.; Du, W.; Zhang, Z.; Bai, P.; Tie, D. Deformation strengthening mechanism of in situ TiC/TC4 alloy nanocomposites produced by selective laser melting. *Compos. Part B Eng.* **2021**, *225*, 109305. [CrossRef]
23. Huo, P.; Zhao, Z.; Du, W.; Bai, P. Deformation and fracture mechanisms of in situ synthesized TiC reinforced TC4 matrix composites produced by selective laser melting. *Ceram. Int.* **2021**, *47*, 19546–19555. [CrossRef]
24. Zhao, Z.Y.; Li, L.; Bai, P.K.; Jin, Y.; Wu, L.Y.; Li, J.; Guan, R.G.; Qu, H.Q. The Heat Treatment Influence on the Microstructure and Hardness of TC4 Titanium Alloy Manufactured via Selective Laser Melting. *Materials* **2018**, *11*, 1318. [CrossRef] [PubMed]

25. Zhao, Z.; Wang, S.; Du, W.; Bai, P.; Zhang, Z.; Wang, L.; Wang, J.; Tie, D. Interfacial structures and strengthening mechanisms of in situ synthesized TiC reinforced Ti6Al4V composites by selective laser melting. *Ceram. Int.* **2021**, *47*, 34127–34136. [CrossRef]
26. Wang, S.; Zhao, Z.; Bai, P.; Du, W.; Li, L. Effect of in situ synthesis TiC on the microstructure of graphene/Ti6Al4V composite fabricated by selective laser melting. *Mater. Lett.* **2021**, *304*, 130715. [CrossRef]
27. Wang, H.; Fang, Z.Z.; Sun, P. A critical review of mechanical properties of powder metallurgy titanium. *Int. J. Powder Metall.* **2010**, *46*, 45–57.
28. Xiao, M.; Colida.; Lai, C.; Dou, E.; Du, L. SLM Formed TC4 Titanium Alloy High Temperature Tensile Mechanical Properties and Fracture Mechanism Analysis. *Appl. Laser* **2021**, *41*, 350–356.
29. Kim, Y.; Lee, J.; Yeom, M.S.; Shin, J.W.; Kim, H.; Cui, Y.; Kysar, J.W.; Hone, J.; Jung, Y.; Jeon, S.; et al. Strengthening effect of single-atomic-layer graphene in metal-graphene nanolayered composites. *Nat. Commun.* **2013**, *4*, 2114. [CrossRef]
30. Zhang, X.; Song, F.; Wei, Z.; Yang, W.; Dai, Z. Microstructural and mechanical characterization of in-situ TiC/Ti titanium matrix composites fabricated by graphene/Ti sintering reaction. *Mater. Sci. Eng. A* **2017**, *705*, 153–159. [CrossRef]
31. Hu, Z.-Y.; Cheng, X.-W.; Li, S.-L.; Zhang, H.-M.; Wang, H.; Zhang, Z.-H.; Wang, F.-C. Investigation on the microstructure, room and high temperature mechanical behaviors and strengthening mechanisms of the (TiB + TiC)/TC4 composites. *J. Alloys Compd.* **2017**, *726*, 240–253. [CrossRef]
32. Baufeld, B.; Brandl, E.; van der Biest, O. Wire based additive layer manufacturing: Comparison of microstructure and mechanical properties of Ti–6Al–4V components fabricated by laser-beam deposition and shaped metal deposition. *J. Mater. Process. Technol.* **2011**, *211*, 1146–1158. [CrossRef]

metals

Article

The Processing Map of Laser Powder Bed Fusion In-Situ Alloying for Controlling the Composition Inhomogeneity of AlCu Alloy

Yang Zhou [1,2], Xiaohan Chen [2], Fan Zhou [2], Xinggang Li [3], Yuhe Huang [3] and Qiang Zhu [2,*]

1 School of Mechatronics Engineering, Harbin Institute of Technology, Harbin 150001, China
2 Department of Mechanical and Energy Engineering, Southern University of Science and Technology, Shenzhen 518055, China
3 Shenzhen Key Laboratory for Additive Manufacturing of High-Performance Materials, Southern University of Science and Technology, Shenzhen 518055, China
* Correspondence: zhuq@sustech.edu.cn

Abstract: In-situ alloying is a facile method for exploring high-performance metallic materials for additive manufacturing. However, composition inhomogeneity is inevitable, and it is a double-edged sword for the properties of in-situ alloyed parts. Appropriately controlling the composition inhomogeneity benefits the applications of in-situ alloying in specific microstructural and properties design. In this work, the Al20Cu alloy was selected as the benchmark alloy to investigate the tailoring of composition inhomogeneity. The morphology and area percentage of composition inhomogeneity in the as-built samples were firstly analyzed. These results provided evidence for the formation of composition inhomogeneity and indicate that its content is tightly dependent on processing parameters. The characteristics of the molten pool under various processing parameters were investigated by modeling the laser remelting process. Based on these, a processing map was established to guide the tailoring of composition inhomogeneity. This study expands the understanding of the formation mechanism of composition inhomogeneity in in-situ alloyed parts and sheds light on employing laser powder bed fusion in-situ alloying for new materials development.

Keywords: in-situ alloying; composition inhomogeneity; computer simulation; processing map

Citation: Zhou, Y.; Chen, X.; Zhou, F.; Li, X.; Huang, Y.; Zhu, Q. The Processing Map of Laser Powder Bed Fusion In-Situ Alloying for Controlling the Composition Inhomogeneity of AlCu Alloy. *Metals* **2023**, *13*, 97. https://doi.org/10.3390/met13010097

Academic Editors: Zhanyong Zhao and Pavel Krakhmalev

Received: 23 November 2022
Revised: 20 December 2022
Accepted: 29 December 2022
Published: 2 January 2023

1. Introduction

Laser powder bed fusion (LPBF) in-situ alloying is a cutting-edge technology, which realizes in-situ alloying of a certain proportion of different alloys or elemental powders by laser during the printing process, providing a simple method for optimizing the microstructure and composition characteristics of components [1]. Modification of alloy composition by in-situ alloying has been widely applied in LPBF [2–7]. However, chemical composition inhomogeneity due to inadequate in-situ alloying has been commonly observed. Generally, the composition inhomogeneity shows a negative impact on the quality of printed parts [5,7–9]. For instance, Lin et al. [6] reported that Si inhomogeneities decreased the ductility of Si-containing FeCoCrNi fabricated by in-situ alloying. However, with proper control, composition inhomogeneity can be utilized to ameliorate the performance of the components [10–13]. T. Zhang et al. [10] found that the strength and toughness of the alloy can be improved synchronously by reasonably utilizing the composition inhomogeneity. Therefore, reasonable control of composition inhomogeneity is a key challenge in manufacturing reliable parts by in-situ alloying.

Composition inhomogeneity is produced through inadequate homogenization of different melt in the molten pool during LPBF in-situ alloying [10,14] in which the melt has undergone an intense mixing process due to Marangoni fluid flow [15]. Although Marangoni fluid flow tends to homogenize the melt in the molten pool, the chaotic mixing traces are preserved due to the small size and rapid cooling of the molten pool of

LPBF [14,16]. Thus, the morphology that arises after the laser melting can be observed in various in-situ alloyed alloys [6,10,17]. Considering the shape and melt flow of the molten pool are tightly connected with the processing parameters of LPBF [18], it is, therefore, a direct way to control the intermixing process by adjusting processing parameters. In the past a few years, considerable research has proved that composition inhomogeneity is alleviated by increasing laser energy input [17,19,20]. However, to our knowledge, no attempt has been made to define a function for processing parameters to characterize the mixing capacity of the corresponding molten pool, which is important for controlling the composition inhomogeneity. Thus, more basic investigations would be advisable.

Recently, increased attention has focused on Al–Cu alloys due to their high strength, corrosion resistance, and low density [21–23]. Meanwhile, the composition inhomogeneity of Al-Cu binary alloys with high copper content can be distinguished by both the distinct change in microstructure and the difference in contrast via backscattered electron (BSE) imaging [19,24]. Therefore, the Al20Cu alloy was selected as the benchmark alloy of the current research, and the mechanism of controlling the composition inhomogeneity via adjusting processing parameters during in-situ alloying was intensely studied experimentally and numerically. The LPBF processing parameters were well designed to avoid the chaotic keyhole, because the chaotic keyhole mode molten pool results in the generation of excessive porosity [18], which damages the performance of the component. The relationship between melt flow behavior and processing parameters was investigated by modeling the laser molten pool under various processing parameters. Based on this, a processing map was established to guide the drafting of processing parameters in in-situ alloying.

2. Materials and Methods

2.1. Sample Preparation and Microstructural Characterization

The gas-atomized commercial pure Al (99.9%) powder and pure Cu (99.9%) were overall spherical with a powder size distribution range of 15–53 μm. The 80 wt.% of pure Al powder and 20 wt.% of pure Cu powder were weighed and mixed with 180 g agate ball by WAB Turbula mixer (WAB, Moutonz, Switzerland) for 8 h for homogenization. The blended powder was measured by energy dispersive spectroscopy (EDS, Zeiss, Oberkochen, Germany) mapping and showed a uniform distribution, as shown in Figure 1a. The size of blended powder was also measured by Matersizer 300+ EV (Malvern Panalytical, Shanghai, China), as shown in Figure 1b.

The experimental samples were printed by EOS 290 equipment (EOS GmbH, Munich, Germany), which is equipped with a spot diameter of 70 μm laser. The power and scanning rate of the laser is the most important processing parameters for the molten pool formation, therefore, in the current work, only these two parameters were studied. The processing parameters used in the current study are listed in Table 1. The processing parameters design refers to the research [23,25] on processing parameters of LPBF aluminum alloy; higher scanning speed was used at high laser power and lower scanning speed at low laser power to avoid the chaotic keyhole and lack-of-fusion pores. The layer thickness and the hatch space should be as small as possible to facilitate the powder melting. The layer thickness was set as 30 μm, slightly larger than the powder average size of 22 μm, and the hatch space was set as 70 μm, equal to the size of the laser spot. A standard alternating x/y-raster scan strategy was used to build the samples, as shown in Figure 1c. The entire printing process was conducted in an inert argon atmosphere.

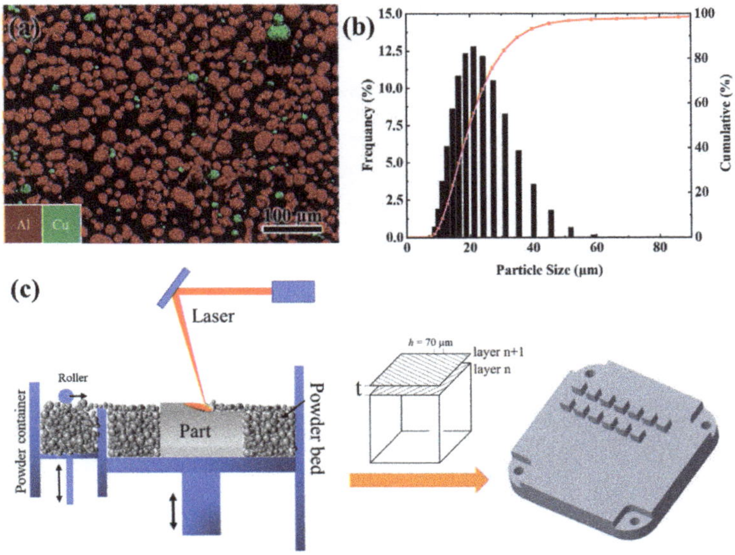

Figure 1. (**a**) Al powder and Cu powder distribution and (**b**) particle size distribution of blended Al20Cu powder. (**c**) Schematic of LPBF process and scanning strategy.

Table 1. Processing parameters used in the sample printing experiment.

Power (W)	Scanning Rate (m/s)	Hatch Distance (μm)	Layer Thickness (μm)
160	0.08	70	30
190	0.15, 0.2, 0.25	70	30
220	0.8, 1.0, 1.2	70	30
250	0.8, 1.2, 1.6	70	30
300	1.2, 1.8	70	30

The specimens were ground and polished to a mirror finish by the conventional metallographic preparation method. Microstructure characterization was conducted on Zeiss Merlin field emission scanning electron microscope (SEM, Zeiss, Oberkochen, Germany) equipped with an energy dispersive spectroscopy (EDS). All characterizations of printed samples were performed on the XY plane (perpendicular to the building direction), which better shows the melting of the powder layer.

2.2. Laser Molten Pool Modeling

Previous studies on LPBF molten pool have revealed that the melt flow and shape of the molten pool is dominated by the Marangoni effect and metal vaporization [26,27]. This study focuses on comparing the effect of processing parameters on the characteristics of the molten pool, such as melt flow and shape of the molten pool. Therefore, the Marangoni effect and metal vaporization are weighed and the powder kinetics are ignored in this model. The LPBF process is simplified as laser remelting of Al20Cu alloy. Figure 2 shows the schematic of computational domain. Only half of the workpiece was taken into account to save computing resources. A movable laser is scanning in the +x direction to melt the metal, and this process is conducted in an argon atmosphere. The interactions of laser and metal, such as melting, solidification, metal evaporation, and Marangoni effect, were included in this model. The melt in the molten pool is treated as incompressible Newtonian laminar flow. This simulation was implemented in the commercial software FLUENT (18.0, ANSYS, Beijing, China). Based on these simplifications, the molten pool characteristics over the shifts in processing parameters were obtained quickly.

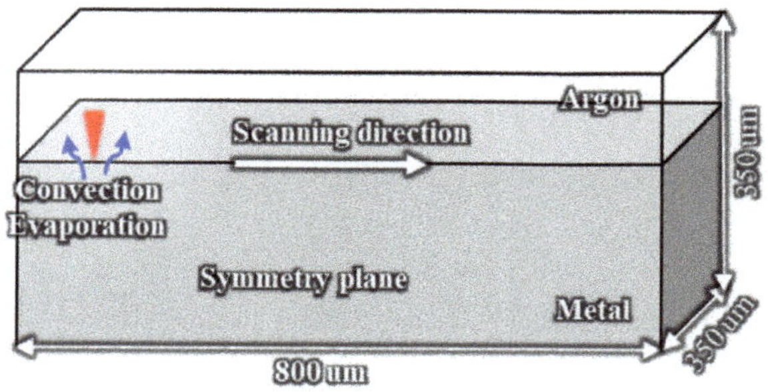

Figure 2. Schematic of boundary condition and computational domain for laser scanning simulation.

The governing conservation equations of mass, momentum, and energy concentration is formulated as follows [27–29]:

$$\frac{\partial \rho}{\partial t} + \nabla \cdot \rho \vec{u} = 0, \tag{1}$$

$$\rho\left(\frac{\partial \vec{u}}{\partial t} + \left(\vec{u} \cdot \nabla\right)\vec{u}\right) = \nabla \cdot [-pI + \mu \nabla \vec{u} + \mu (\nabla \vec{u})^T] + \rho \vec{g} + \boldsymbol{p_s} + \boldsymbol{P_{evp}} + \boldsymbol{F_{SL}}, \tag{2}$$

$$\rho c_p \left(\frac{\partial T}{\partial t} + \vec{u} \nabla T\right) = \nabla \cdot (k \nabla T) + \boldsymbol{Q_l} + \boldsymbol{Q_{con}} + \boldsymbol{Q_{evp}} + \boldsymbol{Q_{SL}}, \tag{3}$$

where $\vec{u}$, ρ, c_p, μ, and k are the fluid velocity, density, specific heat, viscosity, and thermal conductivity, respectively. $\rho \vec{g}$ is the gravitational volume force. The source terms (in bold) were added by coupling with the user-defined functions (UDF), which will be described in detail below.

The gas–liquid interface is captured by the volume-of-fluid (VOF) method and the reactions on the free surface were written as volumetric with a continuum surface force (CSF) method [30].

$$\frac{\partial F}{\partial t} + \nabla \cdot (F \vec{u}) = 0, \tag{4}$$

where F is the volume fraction of gas or metal.

The important reactions in the metal/gas surface are the laser heat flux and metal evaporation in this process. The laser heat flux $\boldsymbol{Q_l}$ input at the metal/gas interface is assumed as a Gaussian laser beam, which can be described as [31]:

$$\boldsymbol{Q_l} = \frac{2P \cdot \eta_l}{\pi r_b^2} exp\left(\frac{-2\left((x - V \cdot t)^2 + y^2\right)}{r_b^2}\right) \cdot |\nabla F| \frac{2\rho_{suf} \cdot c_{Psuf}}{\rho_m c_{Pm} + \rho_g c_{Pg}}, \tag{5}$$

where P is laser power, V is laser scanning rate, r_b is the effective laser beam radius, x and y are the distance to the laser beam's center in x and y directions, η_l is laser absorptivity. $|\nabla F|$ is the absolute value of the gradient of metal phase volume fraction. The specific heat capacity and density of metal, gas, and surface are indicated by c_P and ρ with subscript m, g, and suf, respectively.

The reaction of metal evaporation is included in this model. The vapor recoil pressure $\boldsymbol{P_{vap}}$ and vapor heat loss $\boldsymbol{Q_{vap}}$ can be formulated as [27–29]:

$$\boldsymbol{P_{vap}} = 0.54 P_a exp\left(\frac{M \cdot L_{evp}}{K_B}\left(\frac{T - T_{evp}}{T \cdot T_{evp}}\right)\right) \cdot |\nabla F| \frac{2\rho_{suf}}{\rho_m + \rho_g}, \tag{6}$$

$$\boldsymbol{Q_{vap}} = \frac{-L_{evp} \cdot 0.82}{\sqrt{2\pi RM \cdot T}} \cdot P_a exp\left(\frac{M \cdot L_v}{K_B}\left(\frac{T - T_{evp}}{T \cdot T_{evp}}\right)\right) \cdot |\nabla F| \frac{2\rho_{suf} \cdot c_{Psuf}}{\rho_m c_{Pm} + \rho_g c_{Pg}}, \tag{7}$$

where L_{evp}, T_{evp}, P_a, R, M, and K_B are evaporation latent heat, evaporation temperature, ambient pressure, gas constant, molecular weight, and Boltzmann constant, respectively.

For the non-isothermal free surface, the moment source term $\boldsymbol{p_s}$ caused by surface tension pressure and the Marangoni shear stress was formulated by [28]:

$$\boldsymbol{p_s} = (\gamma\kappa \cdot n - \nabla_s T\frac{d\gamma}{dT}) \cdot |\nabla F| \frac{2\rho_{suf}}{\rho_m + \rho_g}, \tag{8}$$

where κ is the curvature of a free surface, γ is the surface tension, and n is the normal unit vector. $\frac{d\gamma}{dT}$ represent the temperature coefficient of surface tension.

The corresponding boundary condition for heat loss $\boldsymbol{Q_{con}}$ of convection and radiation at the free surface was written as [31]:

$$\boldsymbol{Q_{con}} = -(h_c(T - T_0) + \sigma\varepsilon(T^4 - T_0^4)) \cdot |\nabla F| \frac{2\rho_{suf} \cdot c_{Psuf}}{\rho_m c_{Pm} + \rho_g c_{Pg}}, \tag{9}$$

where h_c is the coefficient of convection heat transfer, σ is Stefan-Boltzmann constant, and ε is the radiant emissivity.

The solid–liquid phase change of metal is considered by set a pseudo solidification interface, which is identified by temperature. The liquid fraction f_L is defined as follows [32]:

$$f_L = \begin{cases} 0, & T < T_S \\ \frac{T - T_S}{T_L - T_S}, & T_S \le T \le T_L, \\ 1, & T_L < T \end{cases} \tag{10}$$

where T_S is the solidus temperature and T_L is the liquidus temperature. Enthalpy-porosity method is used to model the melting and solidification process [32]. The melt velocity is damped when going through the mush zone, and it vanished in the solid. The damping force $\boldsymbol{F_{SL}}$ when fluid goes through the mush zone and the latent heat release or absorb during metal melting or solidification can be expressed as follows:

$$\boldsymbol{F_{SL}} = A_{mush} \frac{(-f_L)^2}{(f_L^3 + 0.001)} \cdot \vec{u}, \tag{11}$$

$$\boldsymbol{Q_{SL}} = -\left[\frac{\partial \rho \Delta L_f}{\partial t} + \nabla \cdot \left(\rho \vec{u} \Delta L_f\right)\right], \tag{12}$$

where ΔL_f is the latent heat of solid-liquid phase change. The parameter A_{mush} is the mushy zone constant, which is set to 1×10^{10} in this model.

The mesh cell size is 4 μm and time step is 10^{-6} s. The second-order upwind method was used for the spatial discretization of energy and momentum. The pressure implicit with splitting of operators (PISO) algorithm was applied for the velocity-pressure coupling. The thermophysical properties of Al20Cu alloy were calculated by Thermo-calc software and listed in Table 2.

Table 2. Thermophysical properties of the material used in this simulation [33].

Property	Al20Cu
Solid density ρ_s (kg m^{-3})	3140
Liquid density ρ_l (kg m^{-3})	2510
Solidus temperature T_S (K)	821.16
Liquidus temperature T_L (K)	875.7
Solid specific heat c_p (J kg^{-1} K^{-1})	850
Liquid specific heat c_p (J kg^{-1} K^{-1})	1010
Solid thermal conductivity k (W m^{-1} K^{-1})	135.4
Liquid thermal conductivity k (W m^{-1} K^{-1})	100

Table 2. *Cont.*

Property	Al20Cu
Heat transfer coefficient h_c (W m^{-2} K^{-1})	100
Radiant emissivity ε	0.2
Temperature coefficient of surface tension $d\gamma/dT$ (N m^{-1} K^{-1})	−0.00035
Latent heat of fusion L_f (J kg^{-1})	3.87×10^5
Latent heat of evaporation L_{evp} (J kg^{-1})	9.46×10^6
Vaporization temperature T_{evp} (K)	2792.15

3. Results and Discussion

3.1. Composition Inhomogeneity of Al20Cu Alloy

Figure 3a shows the composition inhomogeneity morphology of the Al20Cu samples in-situ alloyed at different processing parameters. Figure 3b shows the elements distributed on the dark and bright regions of the BSE micrograph that commonly existed in most of the samples. By comparing the results of the BSE micrograph with the EDS mapping result, the bright area, dark area, and grey area in the BSE micrograph can be distinguished as Cu-rich zone, Cu-poor zone, and sufficient alloyed zone, respectively. It can be clearly observed that the morphology of the composition inhomogeneity within all samples resembled the melt flow morphology, which is an obvious morphology that arose after the laser melting [14]. This morphology is consistent with previous studies on in-situ alloying of Al-based, Ti-based, and high-entropy alloys [2,5,6,10,17]. The composition inhomogeneity regions are reduced with the increase of laser power (marked with the blue dash box in Figure 3a) or the decrease of scanning rate (marked with the red dash box in Figure 3a). A significant reduction in composition inhomogeneity was observed at the hyper-low scanning speed (marked with the black box in Figure 3a).

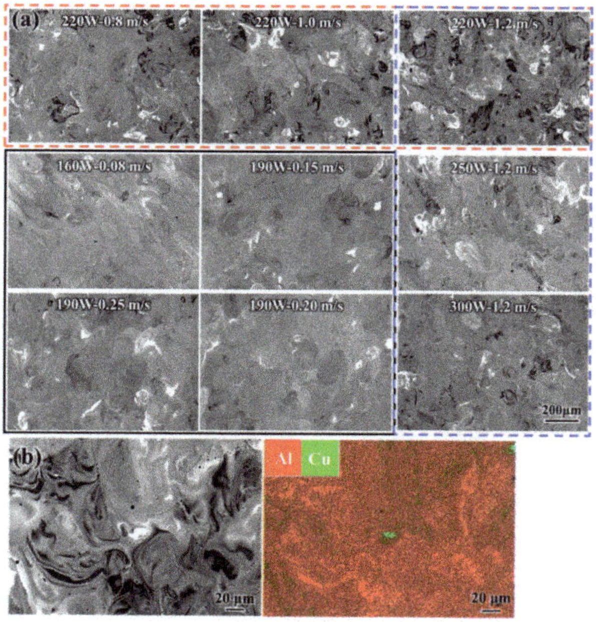

Figure 3. Composition inhomogeneity within the Al-20Cu samples printed at various scanning rates and laser powers: (**a**) BSE micrographs of all samples, and (**b**) EDS mapping analysis of composition inhomogeneity region within the sample processed by laser power of 220 W and scanning rate of 1.2 m/s.

To quantify the extent of alloying, the area percentages of the composition inhomogeneity region in each sample were measured. An example of the measurement process of the composition inhomogeneity region is shown in Figure 4a. Three low magnitudes BSE micrographs in random positions were captured for each sample in the cross-section. The Cu-poor zone and Cu-rich zone in the BSE micrograph are selected by thresholding and quantified by ImageJ software after noise reduction and binarization. The area percentage of the composition inhomogeneity region is defined as the sum of the Cu-poor zone and Cu-rich zone. The normalized enthalpy $\Delta H/h_s$, a most commonly used form, was used to evaluate the laser energy absorbed by a material in each process parameter [34–36]. This number is formulated as follows [35]:

$$\frac{\Delta H}{h_s} = \frac{\eta_l P}{\rho h_s \sqrt{\pi\alpha \cdot r_b^3 \cdot V}}, \quad (13)$$

where h_s, η_l, P, V, α, and r_b are the enthalpy of fusion, laser absorptivity, laser power, laser scanning rate, thermal diffusion coefficient, and laser beam radius, respectively.

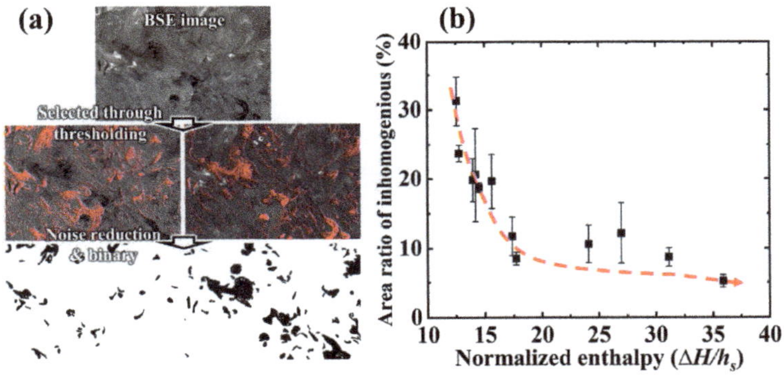

Figure 4. Quantitative characterization of composition inhomogeneity defects within the samples printed by different energy input: (**a**) schematic of measurement process, (**b**) statistical results.

The statistical results show that the area ratio of the composition inhomogeneity region within the sample is closely related to the absorbed laser energy, as shown in Figure 4b. As $\Delta H/h_s$ increased, the area percentage of the composition inhomogeneity region decreases rapidly from 31.3% to 10% and then slowly declined to 5.2%. These results indicated that the composition inhomogeneity can be tailored by adjusting the laser power and scanning rate.

Figure 5 shows the representative microstructure of in-situ alloyed Al20Cu alloy. The fully alloyed Al20Cu alloy exhibits nano-sized cellular α-Al, which is decorated with a bright Al_2Cu phase in the core of the molten pool, while the micro-sized cellular is dispersed along the molten pool boundary (Figure 5b). This featured microstructure is attributed to the rapid cooling of the molten pool [37]. The obvious solidification microstructures were detected in both the Cu-poor zone (Figure 5c) and the Cu-rich zone (Figure 5d) in which granular Al_2Cu and obvious copper dendrites can be, respectively, observed. The composition fluctuations in both regions were analyzed by EDS line analysis (Figure 5c,d). In the Cu-rich zone, the content of copper element steadily rises in the diffusion layer, showing a large fluctuation amplitude but a small fluctuation frequency of composition as a whole. On the contrary, the fluctuation amplitude of composition in the Cu-poor zone is small, while the fluctuation frequency remains high. Beyond the insufficient alloyed zone, the mass percentage of Cu and Al concentration is close to the designed concentration. The violent fluctuation of element content indicates that aluminum and copper elements undergo a violent mixing process in the molten pool. Yet, the extremely high cooling rate of the molten pool preserved this morphology features [10], which also indicates that the life period of the molten pool is important for the formation of composition inhomogeneity.

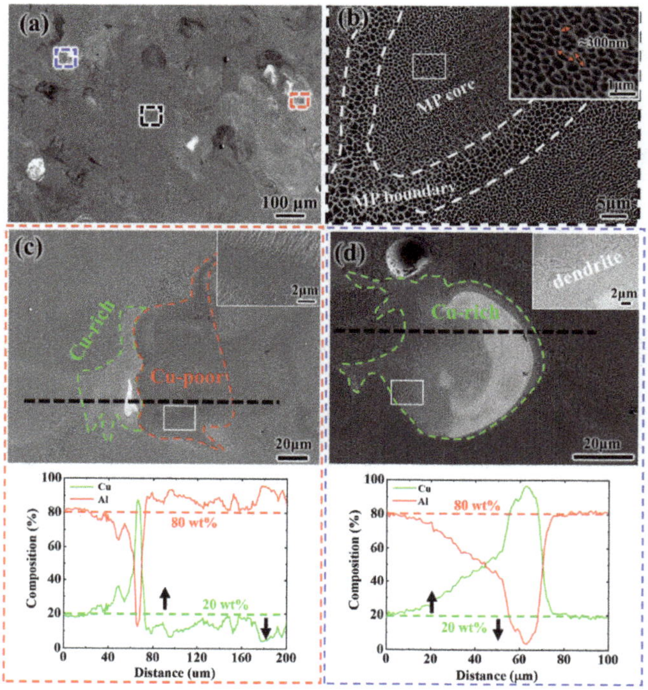

Figure 5. The microstructure of the cross-plane of an as-printed Al20Cu sample: (**a**) a representative SEM micrograph and an enlarged detail of (**b**) fully alloyed zone, (**c**) Cu-poor zone, and (**d**) Cu-rich zone. The EDS line analysis performed across the Cu-poor zone and Cu-rich zone is placed at the bottom of (**c**,**d**).

The above results verify that morphology of composition inhomogeneity is tightly related to the processing parameters. In addition, the present results also indicate the characteristics of the molten pool, such as the melt flow and life period of the molten pool, which is critical in controlling the composition inhomogeneity of the LPBF in-situ alloyed components.

3.2. Characteristics of the Molten Pool under Various Processing Parameters

In the present work, the complicated LPBF process was reduced to a laser re-melting process. Based on this, the trend of molten pool characteristics over the shifts in processing parameters can be obtained quickly. The temperature and velocity field of molten pools in various processing parameters are shown in Figure 6. In all of these molten pools, the high-temperature melt at the gas-metal interface was derived to the low-temperature region, which is a typical Marangoni flow pattern. However, the shape and size of the molten pool changed significantly with the variation of processing parameters. A significant reduction in the size of the molten pool was observed with the increase of scanning rate (marked with the red dash box in Figure 6) or the decrease of laser power (marked with the blue dash box in Figure 6). The shape of molten pool changes from long and elliptical to roundness, when a hyper-low scanning rate is used.

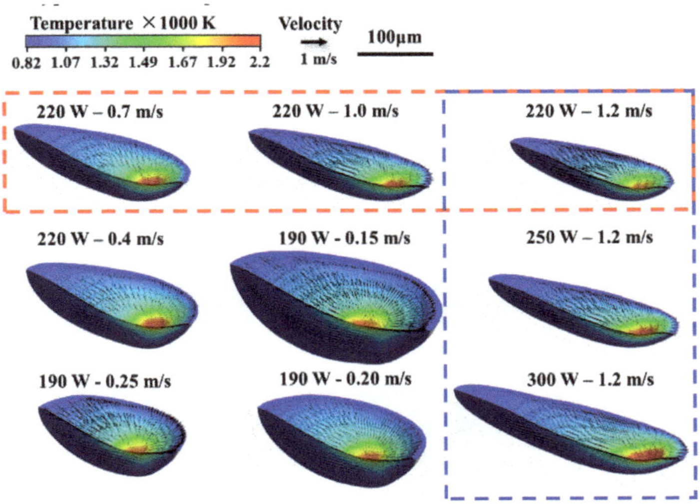

Figure 6. Temperature and velocity field of molten pool under different processing parameters.

The Peclet number P_e, which is the ratio of diffusion rate driven by concentration gradient to convection rate, is used to describe the relative importance of convection and diffusion.

$$P_e = \frac{uL}{D}, \tag{14}$$

where u is the characteristic velocity, L is the characteristic length, and D is the diffusion rate of Cu in the Al matrix (3×10^{-9} m^2/s). The most commonly used form to measure the characteristic length of molten pool in literature is the apparent radius or depth of the molten pool [15,38,39]. However, the molten pool shows a complex geometry shape (Figure 6), in consequence, the geometric characteristics of the molten pool cannot be described basically by the apparent radius or depth of the molten pool. In this work, the molten pool was regarded as a rectangular flow channel with varying cross-sections, and the characteristic length of each cross-section is calculated by hydraulic diameter, as illustrated in Figure 7. The characteristic length L can be calculated as follows:

$$L = \frac{4A}{C_w} = \frac{2hd}{h+d}, \tag{15}$$

where A is the cross-section area of the channel, C_w is the wetting perimeter of the flow channel; h and d are the height and width of the channel, respectively. For each molten pool, six sections are intercepted at equal distances (Figure 7) and the P_e in each plane was calculated. In the present work, the diffusion rate of elements is set to a constant in the *Pe* numbers calculation. Therefore, the *Pe* numbers also reflect the convection strength within the molten pool.

Figure 8 shows the characteristic of the molten pool under various processing parameters. As shown in Figure 8a, the *Pe* number increased rapidly from 8560 to 13,700 and then fluctuated at approximately 13,700 with the increase of normalized enthalpy. It should be noted that the *Pe* number is increased with the increase of the laser power (red ball) or the decrease of the scanning rate (green star) when the normalized enthalpy is lower than 22.5 but does not further improve with the decrease of scanning rate (blue star) when the normalized enthalpy is higher than 22.5. This result indicates that the convection strength cannot be increased unlimitedly by increasing the laser power and decreasing the scanning rate. In addition, the *Pe* is in the order of 10^3, so the mass transfer process within the molten pool was dominated by convection. This result coincides with the experiment that the morphology of the composition inhomogeneity region resembled the convective motion of the melt flow (Figure 3a). Figure 8b,c plot the characteristic length of the molten pool vs. the normalized enthalpy. The average characteristic length is calculated from the average of six channels under each process parameter, and the life period of the molten pool (t_{life}) is the time for the metal to maintain the molten state. Unlike the trends from data in Figure 8a, the life period and dimension of the molten pool is monotonically increasing with the increase of normalized enthalpy, even under the processing parameters of a highly low scanning rate (blue star).

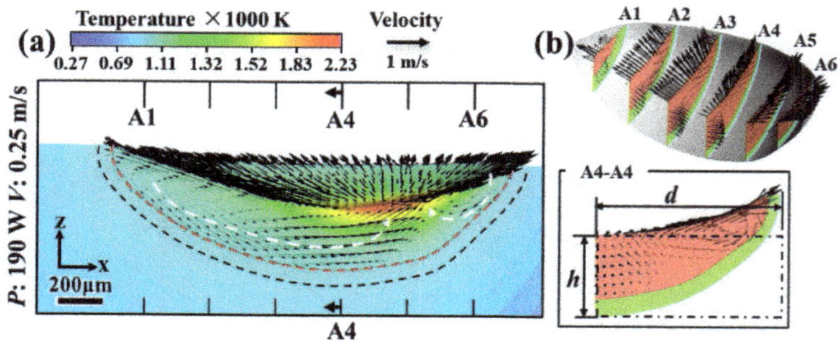

Figure 7. Analysis of convection diffusion by Peclet number Pe: (**a**) Temperature and velocity field of molten pool with laser power of 190 W and scanning rate of 0.25 m/s, and (**b**) the measuring sketch of characteristic length L.

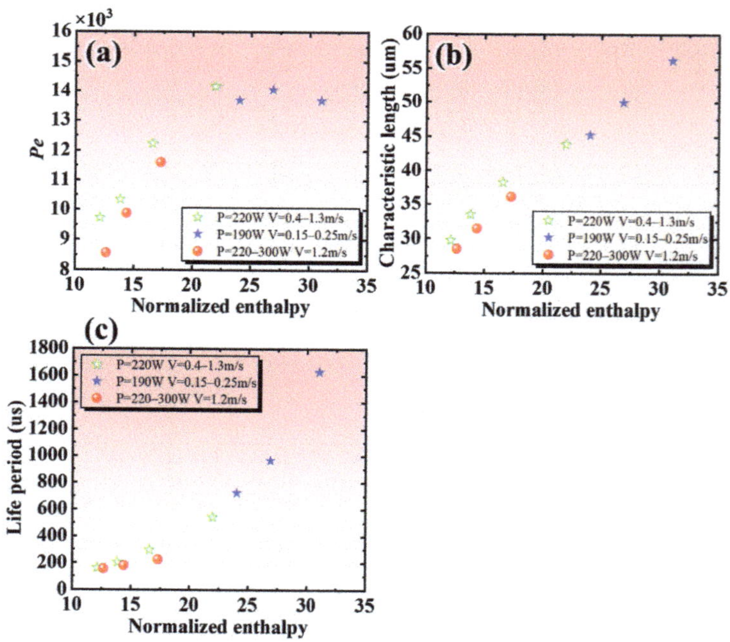

Figure 8. The characteristics of molten pool at various scanning rates and laser powers: (**a**) Pe; (**b**) characteristic length; (**c**) life period.

The above results reveal that the convection strength, time, and dimension of the molten pool are closely related to the normalized enthalpy. Once the normalized enthalpy exceeded a threshold value of 22.5, the convection within the molten pool reaches the peak of intensity in present Al20Cu alloy. As the normalized enthalpy further increases, although the convection strength does not increase, the size of the melt pool and its life period continue to increase, which means that the heterogeneous melts have more time and more space to mix under these conditions. Therefore, further increase of normalized enthalpy can still help to mix the heterogeneous melts. The experiment results present in Figure 3a also show the lower composition inhomogeneity of in-situ alloyed Al20Cu fabricated by the processing parameters with normalized enthalpy exceeding 22.5. Although the complex physical phenomena of in-situ alloying have not been fully revealed, the basic trends of the effect of the adjustment of processing parameters on the pool characteristics can be obtained from the simulation results. As the current work focuses on the influences of processing parameters, a more elaborate simulation will be done in the future.

3.3. The Processing Map for Controlling the Composition Inhomogeneity

The above experimental results have proved the feasibility of tailoring the composition inhomogeneity by adjusting processing parameters (Figure 3). The influence of process parameter adjustment on the factors, which dominate the intermix of heterogeneous melt has been revealed by molten pool simulation (Section 3.2). Here, we shall attempt to summarize a parameter from the simulation results to characterize the mixing ability of the processing parameters to the heterogeneous melts.

The simulation results in Figure 6 show that the melt within the molten pool is a circulating flow. This suggests that the time required for the melt to move around the molten pool (t_{cy}) can be defined as the quotient of molten pool size and melt flow velocity. The molten pool size and melt flow velocity are characterized by the characteristic length L and the characteristic velocity u, respectively, then the parameter t_{cy} is formulated as $t_{cy} = L/u$. It should be noted that this parameter reflects the flow characteristics of the melt in space but does not include the time characteristic. To account for this, a non-dimensional number we called convective stirring number (CSN) is defined as the quotient of the life period of the molten pool (t_{life}) and the time required for the melt to move around the molten pool (t_{cy}), which is expressed as follows:

$$CSN = t_{life} \cdot u/L, \quad (16)$$

Figure 9 plots the variations of convective stirring number with the normalized enthalpy. The convective stirring number is increased with the increase of the laser power (red ball) or the decrease of the scanning rate (green star and blue star), which coincides with the experimental results in Figure 3a. This trend is also consistent with previous studies on the in-situ alloyed Al alloy and high entropy alloy [17,19].

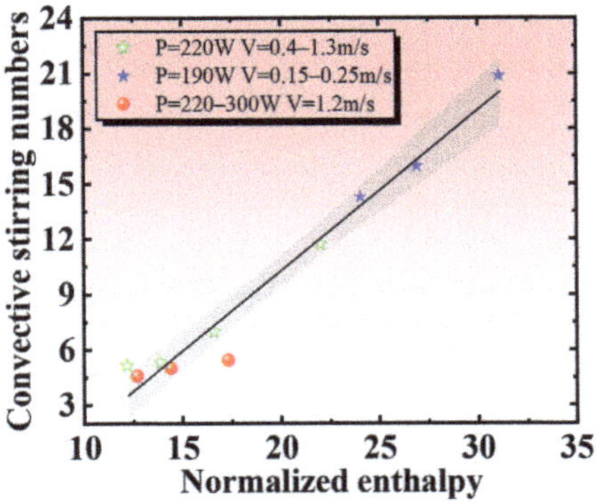

Figure 9. The convective stirring numbers of the molten pool at various scanning rates and laser powers.

In addition, the data in Figure 9 shows that the number of convective stirrings has a good linear fit with the energy input with a R^2 = 0.96. Therefore, the following function can be used to determine the relationship between the convective stirring number and laser parameters:

$$CSN = k \cdot \frac{\eta P}{\rho h_s \sqrt{\pi \alpha r^3 \cdot r_b}} + m, \quad (17)$$

where $k = 0.87$ and $m = 7.1$ are the slope and intercept of the function obtained by linearly fitting the data in Figure 9, respectively. The processing map of CSN is shown in Figure 10. The micrograph of the samples was used to verify the accuracy of the model, which is placed around the processing map. Obviously, the composition inhomogeneity within the sample decreases significantly with increasing convective stirring numbers, which verifies the reliability of the processing map. The processing map in Figure 10a provides that a relatively high CSN can be obtained at the combination of a high laser power and a low scanning rate. However, the excessive input of laser energy will cause a keyhole, which results in the generation of excessive porosity [18], so the low scanning rate should be matched with lower laser power to avoid the chaotic keyhole [23,25].

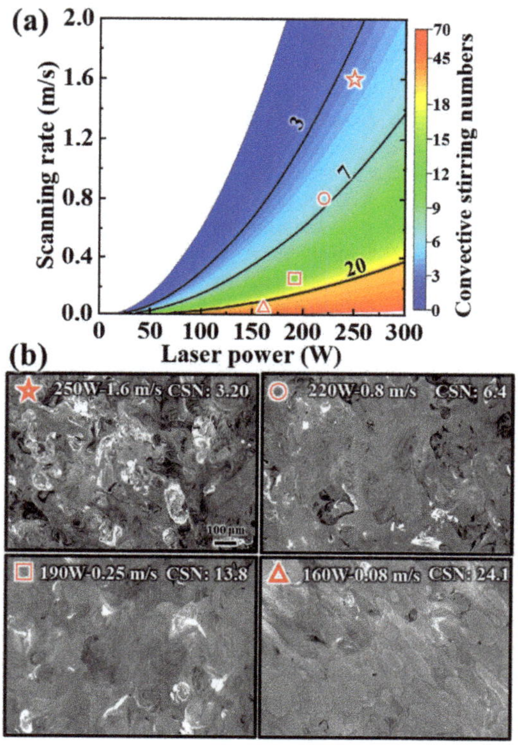

Figure 10. Processing map and experiment results of in-situ alloyed Al20Cu: (**a**) processing map shows the convection stirring number as a function of laser power and scanning rate. (**b**) BSE micrograph of the samples printed at various scanning rates and laser powers.

In addition to adjusting the processing parameters, remelting is a more direct method to increase the convective stirring number. The convective stirring numbers are doubled for each remelting with the same laser parameters. Figure 11 shows the BSE micrograph of samples fabricated by adjusting processing parameters and remelting. The parameters 190 W-0.15 m/s and 220 W-1.3 m/s, with CSN equal to 19.9 and 3.5, respectively, were chosen as a benchmark for sample printing (Figure 11a,c). After once remelting, the CSN of these two parameters is increased to 39.8 and 7.0, respectively (Figure 11b,d). For comparing, the samples fabricated at a processing parameters 400 W-0.22 m/s (CSN = 39.9) and

220 W-0.73 m/s (CSN = 7.1) are shown in Figure 11e,f. For the parameters 190 W-0.15 m/s (Figure 11c), the Al and Cu melt experienced sufficient convective stirring after once remelting, so that the composition inhomogeneity disappeared. However, such a high CSN exists in the area with extremely low scanning speed in the process space, and the sample cannot even be formed due to overheat (Figure 11e). For the parameters 220 W-1.3 m/s with remelting (Figure 11d), the composition inhomogeneity region still exists due to the insufficient convective stirring numbers, and this is consistent with the morphology within the un-remelted sample with the number of convective stirrings equal to 7.1 (Figure 11f). These results showed that the availability of remelting on tailoring the composition inhomogeneity, and the homogeneous Al20Cu samples can be printed when CSN is equal to 39.8.

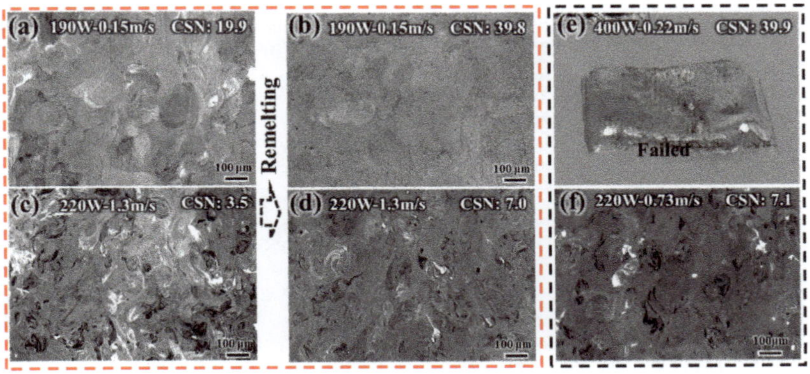

Figure 11. The composition inhomogeneity within the samples to increase CSN by (**a**–**d**) remelting and (**e**–**f**) adjusting processing parameters.

4. Conclusions

The Al20Cu alloy was selected as the benchmark alloy to investigate guidance for the tailoring of composition inhomogeneity by adjusting processing parameters. The main conclusions can be drawn as follows:

(1) The composition inhomogeneity of the in-situ alloyed Al20Cu alloy was characterized. The morphology of the composition inhomogeneity resembled the convective motion of the melt, and it is closely related to the processing parameters;

(2) In the present Al20Cu alloy, the convection within the molten pool reaches the peak of intensity once the normalized enthalpy of processing parameters exceeds 22.5, while the size of the melt pool and its life period continue to increase;

(3) A non-dimensional number CSN was provided to reflect the time-space characteristics of molten pool under various processing parameters. A processing map of convective stirring number was proposed for tailoring the composition inhomogeneity. An Al20Cu sample without composition inhomogeneity was successfully fabricated with the guidance of this processing map.

In summary, the results of this work provide process guidance for tailoring the composition inhomogeneity via LPBF in-situ alloying and contribute to the development of new materials for additive manufacturing by in-situ alloying.

Author Contributions: Y.Z.: conceptualization, methodology, data curation, writing—original draft, writing—reviewing and editing. X.C.: data curation, writing—original draft. F.Z.: investigation. X.L.: writing—original draft preparation. Y.H.: writing—original draft preparation. Q.Z.: writing—reviewing and editing, project administration, supervision. All authors have read and agreed to the published version of the manuscript.

Funding: This research was funded by Shenzhen Science and Technology Innovation Committee, Shenzhen, China (Project Nos. KQTD20170328154443162, JCYJ20210324104610029, JCYJ20210324104608023 and JSGG20210802154210032), the National Natural Science Foundation of China (No. 52074157) and GuangDong Basic and Applied Basic Research Foundation (No. 2021A1515110348).

Data Availability Statement: The data presented in this study are available upon request from the corresponding author.

Acknowledgments: We acknowledge software support from the Center for Computational Science and Engineering at Southern University of Science and Technology, Shenzhen, China.

Conflicts of Interest: The authors declare no conflict of interest.

References

1. Sing, S.L.; Huang, S.; Goh, G.D.; Goh, G.L.; Tey, C.F.; Tan, J.H.K.; Yeong, W.Y. Emerging Metallic Systems for Additive Manufacturing: In-Situ Alloying and Multi-Metal Processing in Laser Powder Bed Fusion. *Prog. Mater. Sci.* **2021**, *119*, 100795. [CrossRef]
2. Yadroitsev, I.; Krakhmalev, P.; Yadroitsava, I. Titanium Alloys Manufactured by In Situ Alloying During Laser Powder Bed Fusion. *Jom* **2017**, *69*, 2725–2730. [CrossRef]
3. Bosio, F.; Manfredi, D.; Lombardi, M. Homogenization of an Al Alloy Processed by Laser Powder Bed Fusion In-Situ Alloying. *J. Alloys Compd.* **2022**, *904*, 164079. [CrossRef]
4. Aboulkhair, N.T.; Simonelli, M.; Parry, L.; Ashcroft, I.; Tuck, C.; Hague, R. 3D Printing of Aluminium Alloys: Additive Manufacturing of Aluminium Alloys Using Selective Laser Melting. *Prog. Mater. Sci.* **2019**, *106*, 100578. [CrossRef]
5. Bosio, F.; Fino, P.; Manfredi, D.; Lombardi, M. Strengthening Strategies for an Al Alloy Processed by In-Situ Alloying during Laser Powder Bed Fusion. *Mater. Des.* **2021**, *212*, 110247. [CrossRef]
6. Lin, D.; Xu, L.; Li, X.; Jing, H.; Qin, G.; Pang, H.; Minami, F. A Si-Containing FeCoCrNi High-Entropy Alloy with High Strength and Ductility Synthesized in Situ via Selective Laser Melting. *Addit. Manuf.* **2020**, *35*, 101340. [CrossRef]
7. Großwendt, F.; Röttger, A.; Strauch, A.; Chehreh, A.; Uhlenwinkel, V.; Fechte-Heinen, R.; Walther, F.; Weber, S.; Theisen, W. Additive Manufacturing of a Carbon-Martensitic Hot-Work Tool Steel Using a Powder Mixture–Microstructure, Post-Processing, Mechanical Properties. *Mater. Sci. Eng. A* **2021**, *827*, 142038. [CrossRef]
8. Huang, S.; Narayan, R.L.; Tan, J.H.K.; Sing, S.L.; Yeong, W.Y. Resolving the Porosity-Unmelted Inclusion Dilemma during in-Situ Alloying of Ti34Nb via Laser Powder Bed Fusion. *Acta Mater.* **2021**, *204*, 116522. [CrossRef]
9. Martinez, R.; Todd, I.; Mumtaz, K. In Situ Alloying of Elemental Al-Cu12 Feedstock Using Selective Laser Melting. *Virtual Phys. Prototyp.* **2019**, *14*, 242–252. [CrossRef]
10. Zhang, T.; Huang, Z.; Yang, T.; Kong, H.; Luan, J.; Wang, A.; Wang, D.; Kuo, W.; Wang, Y.; Liu, C.-T. In Situ Design of Advanced Titanium Alloy with Concentration Modulations by Additive Manufacturing. *Science* **2021**, *374*, 478–482. [CrossRef]
11. Zafari, A.; Lui, E.W.; Jin, S.; Li, M.; Molla, T.T.; Sha, G.; Xia, K. Hybridisation of Microstructures from Three Classes of Titanium Alloys. *Mater. Sci. Eng. A* **2020**, *788*, 139572. [CrossRef]
12. Zafari, A.; Xia, K. Superior Titanium from Hybridised Microstructures–A New Strategy for Future Alloys. *Scr. Mater.* **2019**, *173*, 61–65. [CrossRef]
13. Zafari, A.; Chandran, P.; Xia, K. Superior Tensile Properties in Additively Manufactured Ti Alloys. *Aust. J. Mech. Eng.* **2021**, *19*, 602–608. [CrossRef]
14. Huang, W.; Wang, H.; Rinker, T.; Tan, W. Investigation of Metal Mixing in Laser Keyhole Welding of Dissimilar Metals. *Mater. Des.* **2020**, *195*, 109056. [CrossRef]
15. Guo, Q.; Zhao, C.; Qu, M.; Xiong, L.; Hojjatzadeh, S.M.H.; Escano, L.I.; Parab, N.D.; Fezzaa, K.; Sun, T.; Chen, L. In-Situ Full-Field Mapping of Melt Flow Dynamics in Laser Metal Additive Manufacturing. *Addit. Manuf.* **2020**, *31*, 100939. [CrossRef]
16. Cunningham, R.; Zhao, C.; Parab, N.; Kantzos, C.; Pauza, J.; Fezzaa, K.; Sun, T.; Rollett, A.D. Keyhole Threshold and Morphology in Laser Melting Revealed by Ultrahigh-Speed x-Ray Imaging. *Science* **2019**, *363*, 849–852. [CrossRef]
17. Chen, P.; Yao, X.; Attallah, M.M.; Yan, M. In-Situ Alloyed CoCrFeMnNi High Entropy Alloy: Microstructural Development in Laser Powder Bed Fusion. *J. Mater. Sci. Technol.* **2022**, *123*, 123–135. [CrossRef]
18. Gan, Z.; Kafka, O.L.; Parab, N.; Zhao, C.; Fang, L.; Heinonen, O.; Sun, T.; Liu, W.K. Universal Scaling Laws of Keyhole Stability and Porosity in 3D Printing of Metals. *Nat. Commun.* **2021**, *12*, 1–8. [CrossRef]
19. Skelton, J.M.; Sullivan, E.J.; Fitz-Gerald, J.M.; Floro, J.A. Efficacy of Elemental Mixing of in Situ Alloyed Al-33wt%Cu during Laser Powder Bed Fusion. *J. Mater. Process. Technol.* **2022**, *299*, 117379. [CrossRef]
20. Simonelli, M.; Aboulkhair, N.T.; Cohen, P.; Murray, J.W.; Clare, A.T.; Tuck, C.; Hague, R.J.M. A Comparison of Ti-6Al-4V in-Situ Alloying in Selective Laser Melting Using Simply-Mixed and Satellited Powder Blend Feedstocks. *Mater. Charact.* **2018**, *143*, 118–126. [CrossRef]
21. Wang, P.; Deng, L.; Prashanth, K.G.; Pauly, S.; Eckert, J.; Scudino, S. Microstructure and Mechanical Properties of Al-Cu Alloys Fabricated by Selective Laser Melting of Powder Mixtures. *J. Alloys Compd.* **2018**, *735*, 2263–2266. [CrossRef]

22. Martin, A.; San Sebastian, M.; Gil, E.; Wang, C.Y.; Milenkovic, S.; Pérez-Prado, M.T.; Cepeda-Jiménez, C.M. Effect of the Heat Treatment on the Microstructure and Hardness Evolution of a AlSi10MgCu Alloy Designed for Laser Powder Bed Fusion. *Mater. Sci. Eng. A* **2021**, *819*, 141487. [CrossRef]
23. Tan, Q.; Liu, Y.; Fan, Z.; Zhang, J.; Yin, Y.; Zhang, M.X. Effect of Processing Parameters on the Densification of an Additively Manufactured 2024 Al Alloy. *J. Mater. Sci. Technol.* **2020**, *58*, 34–45. [CrossRef]
24. Gill, S.C.; Kurz, W. Rapidly Solidified AlCu Alloys-I. Experimental Determination of the Microstructure Selection Map. *Acta Metall. Mater.* **1993**, *41*, 3563–3573. [CrossRef]
25. Li, G.; Li, X.; Guo, C.; Zhou, Y.; Tan, Q.; Qu, W.; Li, X.; Hu, X.; Zhang, M.X.; Zhu, Q. Investigation into the Effect of Energy Density on Densification, Surface Roughness and Loss of Alloying Elements of 7075 Aluminium Alloy Processed by Laser Powder Bed Fusion. *Opt. Laser Technol.* **2022**, *147*, 107621. [CrossRef]
26. Aucott, L.; Dong, H.; Mirihanage, W.; Atwood, R.; Kidess, A.; Gao, S.; Wen, S.; Marsden, J.; Feng, S.; Tong, M.; et al. Revealing Internal Flow Behaviour in Arc Welding and Additive Manufacturing of Metals. *Nat. Commun.* **2018**, *9*, 1–7. [CrossRef]
27. Khairallah, S.A.; Anderson, A.T.; Rubenchik, A.; King, W.E. Laser Powder-Bed Fusion Additive Manufacturing: Physics of Complex Melt Flow and Formation Mechanisms of Pores, Spatter, and Denudation Zones. *Acta Mater.* **2016**, *108*, 36–45. [CrossRef]
28. Tseng, C.C.; Li, C.J. Numerical Investigation of Interfacial Dynamics for the Melt Pool of Ti-6Al-4V Powders under a Selective Laser. *Int. J. Heat Mass Transf.* **2019**, *134*, 906–919. [CrossRef]
29. Wang, L.; Zhang, Y.; Yan, W. Evaporation Model for Keyhole Dynamics during Additive Manufacturing of Metal. *Phys. Rev. Appl.* **2020**, *14*, 1. [CrossRef]
30. Brackbill, J.U.; Kothe, D.B.; Zemach, C. A Continuum Method for Modeling Surface Tension. *J. Comput. Phys.* **1992**, *100*, 335–354. [CrossRef]
31. Sun, Z.; Chueh, Y.H.; Li, L. Multiphase Mesoscopic Simulation of Multiple and Functionally Gradient Materials Laser Powder Bed Fusion Additive Manufacturing Processes. *Addit. Manuf.* **2020**, *35*, 101448. [CrossRef]
32. Brent, A.D.; Voller, V.R.; Reid, K.J. Enthalpy-Porosity Technique for Modeling Convection-Diffusion Phase Change: Application to the Melting of a Pure Metal. *Numer. Heat Transf.* **1988**, *13*, 297–318. [CrossRef]
33. Chen, S.; Zhao, Y.; Tian, S.; Gu, Y.; Zhan, X. Study on Keyhole Coupling and Melt Flow Dynamic Behaviors Simulation of 2219 Aluminum Alloy T-Joint during the Dual Laser Beam Bilateral Synchronous Welding. *J. Manuf. Process.* **2020**, *60*, 200–212. [CrossRef]
34. Rubenchik, A.M.; King, W.E.; Wu, S.S. Scaling Laws for the Additive Manufacturing. *J. Mater. Process. Technol.* **2018**, *257*, 234–243. [CrossRef]
35. Hann, D.B.; Iammi, J.; Folkes, J. A Simple Methodology for Predicting Laser-Weld Properties from Material and Laser Parameters. *J. Phys. D. Appl. Phys.* **2011**, *44*, 445401. [CrossRef]
36. Metelkova, J.; Kinds, Y.; Kempen, K.; de Formanoir, C.; Witvrouw, A.; Van Hooreweder, B. On the Influence of Laser Defocusing in Selective Laser Melting of 316L. *Addit. Manuf.* **2018**, *23*, 161–169. [CrossRef]
37. Delahaye, J.; Habraken, A.M.; Mertens, A. Influence of Si precipitates on fracture mechanisms of AlSi10Mg parts processed by Selective Laser Melting. *Acta Mater.* **2019**, *175*, 160–170. [CrossRef]
38. Gan, Z.; Yu, G.; He, X.; Li, S. Numerical Simulation of Thermal Behavior and Multicomponent Mass Transfer in Direct Laser Deposition of Co-Base Alloy on Steel. *Int. J. Heat Mass Transf.* **2017**, *104*, 28–38. [CrossRef]
39. Li, Z.; Yu, G.; He, X.; Tian, C.; Li, S.; Li, H. Probing Thermocapillary Convection and Multisolute Dilution in Laser Welding of Dissimilar Miscible Metals. *Int. J. Therm. Sci.* **2022**, *172*, 107242. [CrossRef]

metals

MDPI

Article

Microstructure and Mechanical Properties of Selective Laser Melted Reduced Activation Ferritic/Martensitic Steel

Zihao Yuan [1,*], Chenyu Xiao [1], Hao Liu [1], Xuejing Sun [1], Guangyi Zhao [1], Hongbin Liao [1], Wei Qian [1], Jingkun Yuan [2], Changshi Lao [2] and Huajie Wang [1,*]

1 Southwestern Institute of Physics, Huangjing Road Number 5, Chengdu 610041, China
2 Additive Manufacturing Institute, College of Mechatronics and Control Engineering, Shenzhen University, Shenzhen 518060, China
* Correspondence: yuanzihao@swip.ac.cn (Z.Y.); wanghuajie@swip.ac.cn (H.W.); Tel.: +86-28-82850482 (Z.Y. & H.W.)

Abstract: Cube and tensile samples of reduced activation ferritic/martensitic steel were formed at different laser powers and scanning velocities using a selective laser melting process; the microstructural characteristics and tensile properties of the cube and tensile samples were investigated in this study. The experimental results showed that the SLMed CLF-1 samples that formed with selected laser melting were near-fully dense, and the relative density of the SLMed CLF-1 samples exceeded 99%. Meanwhile, there were numerous nano-sized spherical and needle-like precipitate dispersions distributed in the grains and boundary of the grains, and the precipitates were mainly composed of M23C6 carbide and MX carbide. The microstructure was composed of columnar grains and equiaxed grains arranged in a sequence, and the smallest average size of the grains was 15 ± 2.1 μm when measured at 320 W of power and 800 mm/s scanning velocity. In addition, the sample at 320 W of power and 800 mm/s scanning velocity exhibited higher yield strength (875 ± 6.0 MPa) and higher elongation (25.6 ± 0.8%) than that of the sample at 200 W of power, 800 mm/s scanning velocity, yield strength of 715 ± 1.5 MPa, and elongation of 22.6 ± 1.2%.

Keywords: CLF-1 steel; selective laser melting; process parameters; microstructure; mechanical properties

Citation: Yuan, Z.; Xiao, C.; Liu, H.; Sun, X.; Zhao, G.; Liao, H.; Qian, W.; Yuan, J.; Lao, C.; Wang, H. Microstructure and Mechanical Properties of Selective Laser Melted Reduced Activation Ferritic/Martensitic Steel. *Metals* **2022**, *12*, 2044. https://doi.org/10.3390/met12122044

Academic Editor: Zhanyong Zhao

Received: 30 October 2022
Accepted: 23 November 2022
Published: 28 November 2022

1. Introduction

In the past decades, reduced activation ferritic/martensitic (RAFM) steel has been widely selected as primary structural materials in the nuclear industry, including test blanket modules (TBMs), nuclear power cladding, and cores, owing to its good advantages of low activation, good radiation resistance, and outstanding mechanical and thermophysical properties [1–4]. The service environment of RAFM steel structure that are used in the nuclear industry usually needs to withstand high thermal load, and it needs to match the plasma configuration [5–8]. Therefore, RAFM steel structures have complex curved surfaces and require high precision. These conditions bring great challenges to traditional processing techniques, and new forming strategies to solve this conflict are desired in the manufacturing of reduced activation ferritic/martensitic steel. Additive manufacturing (AM) can form complex components with high design flexibility because of the layer-by-layer processing methods. The AM approach is mainly divided into laser-directed energy deposition (LDED) and selective laser melting (SLM) according to the powder feeding forms. The LDED technique is widely used in the manufacture and repair of large parts because of the high deposition rate and high degree of freedom of the powder feeding nozzle. The SLM method is suitable for manufacturing precision parts owing to its small molten pool [9–12]; therefore, SLM has been used to confirm the feasibility of producing the RAFM steel component [13–15].

Huang et al. [13] utilized the SLM method to fabricate an RAFM steel component in 2018; while the SLMed RAFM steels have poor formability with many porosities and cracks, Thomas et al. [15] produced RAFM steel walls and solid buildings using SLM. The SLM-built RAFM steel achieved poor anisotropy and lower ductility (7%) compared with the forgings in elongation (~20%) [16]. Huang et al. [14] reported that the defects can be dramatically eliminated by hot isostatic pressing, and as a result the tensile strength of the as-deposited material is improved. However, the use of hot isostatic pressing process causes the tensile strength to deteriorate, and it cannot guarantee the elimination of all the porosities and micro-cracks. Meanwhile, this treatment increases the cost and processing time and contradicts the concept of AM one-time molding. Therefore, it is necessary to systematically study the relationship between the SLMed RAFM steel processing, structure, and performance. In this study, RAFM steels were fabricated using the SLM method under various laser powers and scanning velocities, and the microstructure and tensile properties were investigated using scanning electron microscope (SEM), optical microscopy (OM), and X-ray diffraction (XRD) analyses to further understand the relationship between the processing, microstructure, and properties. The evolution behavior of grain morphology was also revealed in detail based on the experimental results.

2. Materials and Experimental Procedures

2.1. Materials

The RAFM steel raw material CLF-1 was developed by the Southwest Institute of Physics in this experiment. CLF-1 is intended to be used as the first-wall material for fusion reactors in the future. The manufacturing methods for CLF-1′s powder particle are usually one of two methods: gas atomization (GA) and plasma-rotating electrode process (PREP) [17]. The GA approach is the most common method of powder production. Compared with the GA method, the PREP technique has many advantages, such as higher sphericity and lower porosity [17,18]. Considering the characteristics of sphericity and porosity, the powder was produced using the PREP method and was provided by the Institute of Metal Research, Chinese Academy of Sciences. The purpose of this research was to apply RAFM steel in a reactor. The interaction between the neutrons and Mo, Nb, and N elements of the RAFM steel in the reactor will lead to radioactivity in RAFM steel; therefore, the Mo, Nb, and N elements of the RAFM steel would be replaced with W, Ta, and V for the purpose of reducing the activation of RAFM steel. Table 1 lists the chemical compositions of the CLF-1 powder. It can be clearly seen from Table 1 that the CLF-1 steel in this investigate was a low-carbon martensitic steel, and the components of CLF-1 steel were C, Cr, W, Ta, V, Mn, and Fe, and the mass fraction of Fe was 88.16%.

Table 1. Chemical composition of the CLF-1 steel powder.

Element	C	Cr	W	Ta	V	Mn	Fe
Wt%	0.11	9.4	1.5	0.15	0.21	0.47	88.16

Figure 1 shows the scanning electron microscope picture of the CLF-1 powder produced using the PREP method. It can be clearly seen from Figure 1 that the average particle size range of the powder particles is 31.3 μm, and the powder particles have good sphericity. There is no obvious hollow in the powder particles.

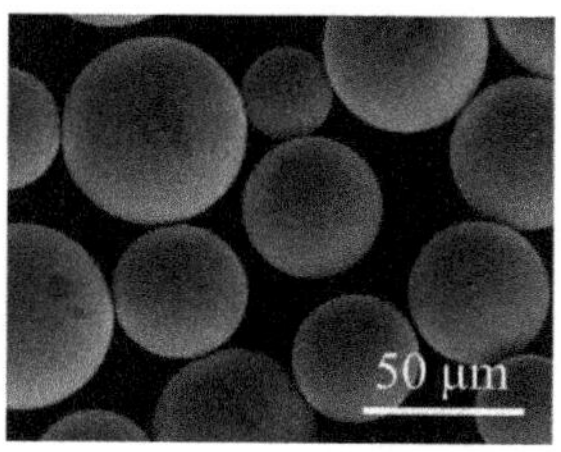

Figure 1. Scanning electron microscope picture of the CLF-1 powder produced using the PREP method.

2.2. Experimental Procedures

RAFM steel powder particles were dried for 2 h at a temperature of 120 °C to remove moisture at the beginning of the experiment. The selective laser melting experiment was carried out in a BLT S310 SLM system. This system contains a 500 W fiber laser and stable powder spreading system. The layer thickness of the powder was set as 30 μm, and the hatch spacing was 85 μm. The laser size was 80 μm. Cuboid samples with dimensions of 10 × 10 × 10 mm^3 were fabricated for microstructural analysis, and plate-shaped tensile specimens were horizontally fabricated (see Figure 2). In order to reduce the stress concentration, the scanning strategy adopted zigzag scanning. The laser powders and the scanning speeds were selected considering the optimized process parameters. Table 2 lists the process parameters used in this experiment, and one cuboid sample was formed for each of the parameter settings. The densities of the SLMed CLF-1 cuboid samples were measured using Archimedes' method with distilled water. The phase analysis of the as-deposited samples was carried out using X-ray diffraction (XRD) with Cu Kα radiation, and the type of XRD used was a Rigaku MiniFlex 600 (Rigaku Inc., Tokyo, Japan); the scanning speed of the XRD analysis was run at about 10 ∘/min. The samples were mechanically ground with #400–2000 before observing the microstructure and polished with 2.5 μm of polishing paste; they were then etched by the solution Vilella's reagent (1 g of picric acid, 10 mL of hydrochloric acid, and 100 mL of ethanol). The microstructure of the sample was examined by optical microscopy using a Keyence VHX-5000 microscope (Keyence Inc., Osaka, Japan) and scanning electron microscope (SEM, FE-SEM, ZEISS MERLIN Compact, Oberkochen, Germany). Hardness tests were carried out using a Struers Duramin-A300 Vickers microhardness tester (Struers Inc., Copenhagen, Denmark) at load of 500 g with 15 s. Tensile tests of the samples were conducted using an INSTRON11-96 electronic testing machine (Instron Inc., High Wycome, England). Three tensile specimens were tested for each condition. The balanced phase diagram and continuous cooling transformation curve (CCT) were calculated using JMatPro software (V5.1, Sente Software Inc., Surrey, England). JMatPro is a software based on material type which has been used in different studies and has good agreement with the predicted material's characteristics (alloy composition, phases, phase composition, physical and mechanical properties, etc.) [12]. The element composition setting in JMatPro, which is used to calculate CCT and the balanced phase diagram, was same as in Table 2. The material type of calculation was selected to be stainless steel. The temperature calculation ranged from 10 °C to 1600 °C, and the calculation step was conducted at 10 °C.

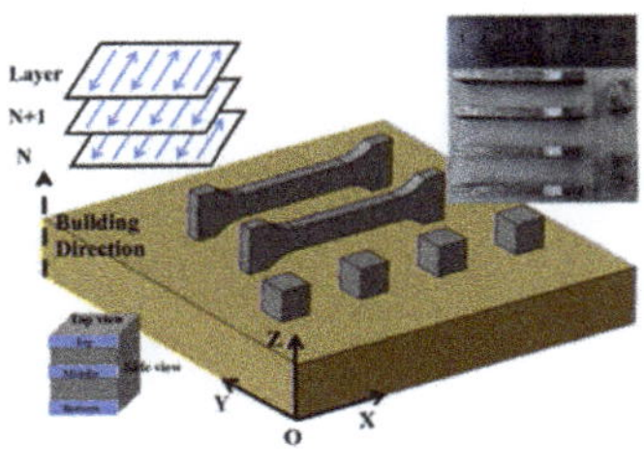

Figure 2. Schematic illustration of the scanning strategies and fabrication of bulks and tensile specimens.

Table 2. The process parameters used in this investigation.

Laser Power (W)	Scanning Velocity (mm/s)	Layer Thickness (μm)	Hatch Spacing (μm)
200	600	30	85
	800		
320	600		
	800		

3. Results and Discussion

3.1. Microstructure

Figures 3 and 4 show the morphologies of the CLF-1 grains from the top view and side view under different process parameters, respectively. As can be seen from Figures 3 and 4, there were no obvious cracks and pores in the samples, and the relative density of all the samples exceeded 99%. The sample reached the highest relative density in the conditions of 320 W power and 800 mm/s scanning velocity, where the relative density was 99.73%. Meanwhile, the microstructure of the grains had a similar morphology; the average grain size was smallest in the conditions of 320 W power and 800 mm/s scanning velocity, and the average grain size was 15 ± 2.1 μm.

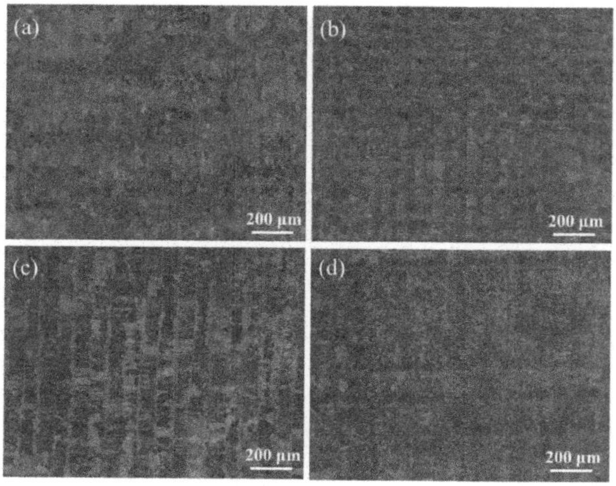

Figure 3. Morphologies of grains under optical microscopy from top view (in XOY section): (**a**) 200 W power and 600 mm/s scanning velocity; (**b**) 200 W power and 800 mm/s scanning velocity; (**c**) 320 W power and 600 mm/s scanning velocity; (**d**) 320 W power and 800 mm/s scanning velocity.

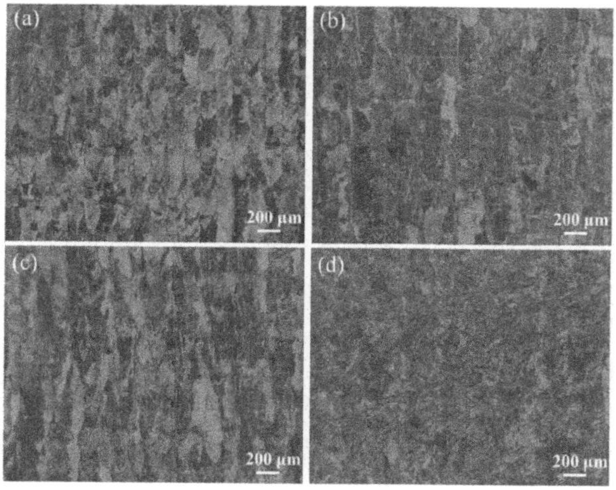

Figure 4. Morphologies of grains under optical microscopy from side view (in XOZ section): (**a**) 200 W power and 600 mm/s scanning velocity; (**b**) 200 W power and 800 mm/s scanning velocity; (**c**) 320 W power and 600 mm/s scanning velocity; (**d**) 320 W power and 800 mm/s scanning velocity.

The growth mode and structure morphology during the solidification process are dependent on the temperature gradient (G) and the growth rate (R). The growth rate (R) can be expressed as [19]:

$$R = V_s \times \cos\theta \quad (1)$$

where V_s is the scanning velocity and θ is the angle between V_s and R. The temperature gradient G can be calculated as follows [20]:

$$G = \frac{2K(T - T_0)^2}{\varepsilon P} \quad (2)$$

where T is the liquid temperature of the alloy, T_0 is the initial temperature of the substrate, ε is the laser absorption coefficient, and P is the laser power. The liquid temperature T of the alloy can be estimated by [21,22]:

$$T(x,y,z) = \frac{1}{2\pi\lambda} \cdot \frac{4P\beta}{\pi D^2} \int_{-\frac{D}{2}}^{\frac{D}{2}} \int_{-\sqrt{\frac{D^2}{4} - x_0{}^2}}^{\sqrt{\frac{D^2}{4} - x_0{}^2}} \frac{\exp[-(\frac{V_S(x-x_0)}{2\alpha} + \frac{V_S\sqrt{(x-x_0)^2+(y-y_0)^2+z^2}}{2\alpha})]}{\sqrt{(x-x_0)^2 + (y-y_0)^2 + z^2}} dxdy \quad (3)$$

where P is the laser power, D is the diameter of the laser beam, and β is the laser absorptivity, which can be calculated by: $\alpha \times c \times \rho$, where α is heat dissipation coefficient, c is the specific heat capacity, and ρ is the density of material.

According to Equation (3), T is proportional to the laser power P; therefore, T increases with an increase in P. According to Equation (2), the temperature gradient G is proportional to liquid temperature T^2 of the alloy; as a result, the temperature gradient G increases with an increase in laser power P. Hence, one conclusion can be deduced that G and R are greatest at 320 W power and 800 mm/s scanning velocity conditions. According to rapid solidification, the grains become smaller as the $G \times R$ increases [23].

It can also be seen from Figures 3 and 4 that columnar grains and equiaxed grains were arranged in sequence along the Y-direction, the equiaxed grains were formed along the center line of the laser track, and epitaxially grown columnar coarse grains penetrated through multiple deposition layers; meanwhile, the width of these columnar grains was about 85 µm, which matched the hatch spacing well. It is worth noting that the direction of epitaxial growth of the columnar grains was not perpendicular to the deposition direction, which is related to the direction of heat diffusion during the solidification of the molten pool. The authors believe that there is a temperature gradient in the molten pool from the center of the molten pool to the outside during the solidification process which, combined with the subsequent laser remelting, resulted in the columnar grains and equiaxed grains being arranged in sequence along the Y-direction.

Figure 5 shows the XRD patterns of the as-deposited CLF-1 steel at 200 W power and 800 mm/s scanning velocity and 320 W power and 800 mm/s scanning velocity, respectively. As can be seen from Figure 5, the diffraction peaks were 44.6°, 65°, 82.3°, which are consistent with the diffraction peaks of ferrite/martensite. Meanwhile, there was no residual austenite detected due to the high cooling rate during the SLM process. It can be seen from Figure 5b–d that the diffraction peak shifted toward the large angle with increasing laser power. The authors believe that as the laser energy density increases, the redistribution of large-sized atoms (Cr, V, and Ta) and interstitial atoms caused the interplanar spacing to decrease, which resulted in this phenomenon.

Figure 6 shows the SEM images of the SLMed CLF-1 steels observed from the top view. It can be seen that numerous nano-sized spherical and needle-like precipitates were dispersion distributed in the grains and boundary of the grains. These precipitates mainly play a role in dispersion strengthening by pinning dislocations and subgrain boundaries to increase the strength [24]. The precipitates of the SLMed CLF-1 steels were mainly as follows: $M_{23}C_6$ carbide, MX carbide, and the laves phase shown in the balanced phase diagram (Figure 7). Because laves phases are often generated during the creep process and have an irregular shape [25], the precipitates were $M_{23}C_6$ carbide and MX carbide.

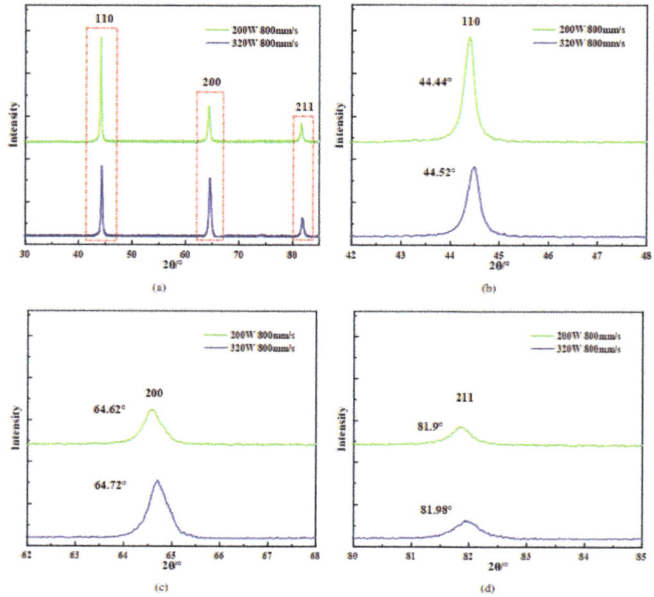

Figure 5. (**a**) XRD patterns of the as-deposited RAFM steels, (**b**) enlargement of the (110) peak, (**c**) enlargement of the (200) peak, and (**d**) enlargement of the (211) peak.

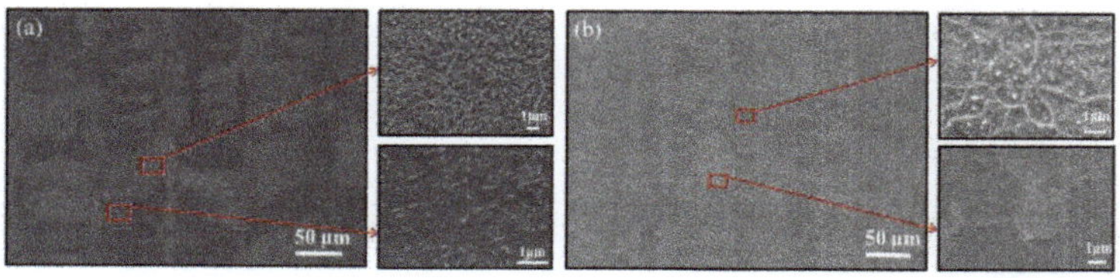

Figure 6. SEM images observed from the top view. (**a**) 200 W power and 800 mm/s scanning velocity; (**b**) 320 W power and 800 mm/s scanning velocity.

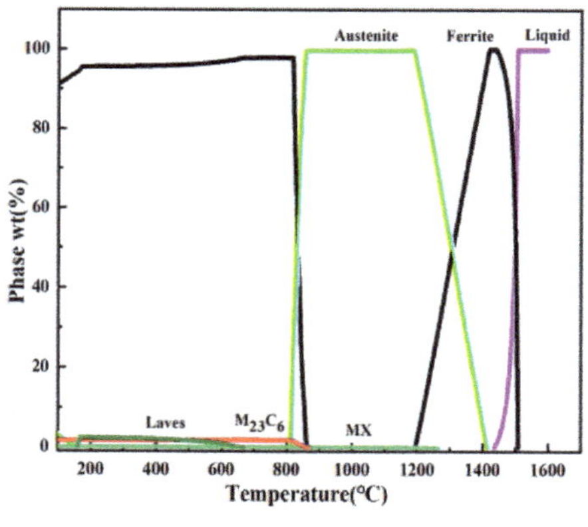

Figure 7. Balanced phase diagram.

3.2. Mechanical Properties

Table 3 lists the micro-hardness of the SLMed CLF-1 steel from the bottom to the top along the deposition direction. As can be seen from Table 3, with the formation of martensites during the SLM process, the micro-hardness became high. It can be seen from Table 3 that the increase in laser power had no significant effect on the micro-hardness. The average micro-hardness of the bottom, middle, and top was 312 ± 7 HV, 289 ± 8 HV, and 293 ± 3 HV at 200 W power with 800 mm/s scanning velocity process, and 323 ± 6 HV, 310 ± 3 HV, and 296 ± 5 HV at 320 W power with 800 mm/s scanning velocity process, respectively. It can also be seen that the micro-hardness from the bottom, middle, and top of the sample was similar, which indicates that the microstructure uniformity was relatively consistent during the SLM Process.

Table 3. Micro-hardness of the SLMed CLF-1 steel.

	Location	200 W 800 mm/s	320 W 800 mm/s
Micro-hardness (HV)	Top	293 ± 3	296 ± 5
	Middle	289 ± 8	310 ± 3
	Bottom	312 ± 7	323 ± 6

Figure 8 shows the stress–strain curves of the SLMed CLF-1 steels. As can be seen from Figure 8, the sample at 320 W power with 800 mm/s scanning velocity exhibited a higher yield strength of 875 ± 6.0 MPa and higher elongation of 25.6 ± 0.8% than that of the sample with 200 W power, 800 mm/s scanning velocity, yield strength of 715 ± 1.5 MPa, and elongation of 22.6 ± 1.2%. The ultimate tensile strength samples at 200 W and 320 W power with 800 mm/s scanning velocity were 710 ± 5.1 MPa and 553 ± 7.2 MPa, respectively. The ductility and strength of the SLMed samples was about the same as the forgings [16]. In polycrystals, the yield strength(σ_s) and the average grain diameter(d) satisfy the Hall–Petch relationship: $\sigma_s = \sigma_0 + Kd^{-1/2}$, where K is the influence coefficient of the grain boundary on deformation, which is related to the grain boundary structure. The higher yield strength at 320 W power and 800 mm/s scanning velocity can be explained by the average grain diameter(d) being smaller than that of the sample with 200 W power and 800 mm/s scanning velocity.

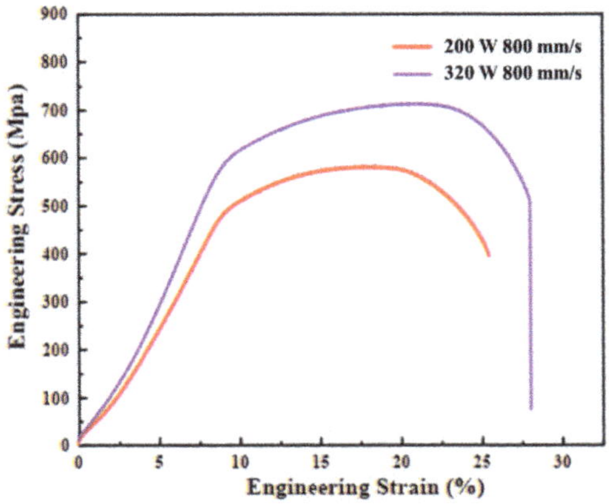

Figure 8. Stress–strain curves of the SLMed CLF-1 steels.

Owing to the unique microstructure formed by the SLM process as shown in Figure 3, Figure 4, and Figure 6, stress concentration occurred at the interface between the columnar grains and the equiaxed grains. The ductility increased due to the stress being dispersed in numerous equiaxed grains. Many fine carbides precipitating from substrate play an important role in precipitation hardening to increase the strength. As a result, the SLMed CLF-1 steel displayed excellent mechanical properties.

4. Conclusions

In this study, the cube and tensile samples of the SLMed CLF-1 steel were formed with the selected laser melting technique at different laser powers and scanning velocities. Furthermore, the microstructure and mechanical properties of these samples have been systematically researched. The conclusions are shown as follows:

(1) The relative density of the SLMed CLF-1 samples exceeded 99%, and there were no obvious tracks and pores in the samples due to using the PREP powder.
(2) Numerous nano-sized spherical and needle-like precipitates were dispersion distributed in the grains and boundary of the grains, and the precipitates were mainly composed of $M_{23}C_6$ carbide and MX carbide. The microstructure was composed of columnar grains and equiaxed grains, and the smallest average size of the grains was 15 ± 2.1 μm under the procedure parameters of 320 W power and 800 mm/s scanning velocity.
(3) The sample at 320 W power and 800 mm/s scanning velocity exhibited higher yield strength (875 ± 6.0 MPa) and higher elongation ($25.6 \pm 0.8\%$) than that of the sample at 200 W power, 800 mm/s scanning velocity, yield strength (715 ± 1.5 MPa), and elongation ($22.6 \pm 1.2\%$).

The investigation showed that the room temperature comprehensive mechanical properties of SLMed CLF-1 steel are same as the forgings. The investigation also provides a new processing strategy for fabricating SLMed CLF-1 steel components. However, the microstructure of SLMed CLF-1 steel has coarse columnar grains, and the obvious anisotropy of these columnar grains will cause significant differences in the tensile properties of different deposition directions. It is necessary to focus on the deposition samples of different directions in the future. Furthermore, SLMed CLF-1 steels have been used in high-temperature environments for a long time, and our follow-up work will study the high-temperature mechanical properties of SLMed CLF-1 steels samples. This work lays the foundation for the application of SLMed CLF-1 steel in the nuclear industry and serves as a significant guide suggestion for the research of first-wall cladding material.

Author Contributions: Z.Y.: Conceptualization, methodology, writing—original, draft; C.X.: investigation, methodology, data curation; H.L. (Hao Liu): methodology, writing—original; X.S.: investigation, draft, writing—review and editing; G.Z.: writing—review and editing; H.L. (Hongbin Liao): formal analysis; W.Q.: conceptualization; J.Y.: methodology; C.L.: supervision; H.W.: investigation, resources, writing—review and editing. All authors have read and agreed to the published version of the manuscript.

Funding: This work was supported by the National Magnetic Confinement Fusion Science Program of China (Project No. 2019YFE03090100).

Data Availability Statement: The data that support the findings of this study are available from the corresponding author, upon reasonable request.

Conflicts of Interest: The authors declare that they have no conflict of interest in this work.

References

1. Zinkle, S.J.; Busby, J.T. Structural materials for fission & fusion energy. *Mater. Today* **2009**, *12*, 12–19.
2. Rietema, C.J.; Walker, M.A.; Jacobs, T.R.; Clarke, A.J.; Clarke, K.D. High-throughput nitride and interstitial nitrogen analysis in ferritic/martensitic steels via time-of-flight secondary ion mass spectrometry. *Mater. Charact.* **2021**, *179*, 111357. [CrossRef]
3. Odette, G.; Alinger, M.; Wirth, B. Recent developments in irradiation-resistant steels. *Annu. Rev. Mater. Res.* **2008**, *38*, 471–503. [CrossRef]
4. Xia, Z.; Zhang, C.; Liu, W.; Yang, Z. Effect of surface mechanical attrition treatment on precipitation behaviors and ferrite grains in heat resisting steels. *ISIJ Int.* **2014**, *54*, 1935–1942. [CrossRef]
5. Shin, S.H. A Study on Corrosion Behaviors of Ferritic/Martensitic Steels in Liquid Sodium Condition for Sodium-cooled Fast Reactor Application. Ph.D. Thesis, Graduate School of Unist, Ulsan, Republic of Korea, 2015.

6. Hirai, T.; Barabash, V.; Escourbiac, F.; Durocher, A.; Ferrand, L.; Komarov, V.; Merola, M. Iter divertor materials and manufacturing challenges. *Fusion Eng. Des.* **2017**, *125*, 250–255. [CrossRef]
7. Tanigawa, H.; Shiba, K.; Sakasegawa, H.; Hirose, T.; Jitsukawa, S. Technical issues related to the development of reduced-activation ferritic/martensitic steels as structural materials for a fusion blanket system. *Fusion Eng. Des.* **2011**, *86*, 2549–2552. [CrossRef]
8. Feng, K.; Wang, X.; Feng, Y.; Chen, Y.; Zhao, Z.; Li, Z.; Wang, P.; Wang, Q.; Ye, X.; Zhao, F.; et al. Current progress of chinese hccb tbm program. *Fusion Eng. Des.* **2016**, *109*, 729–735. [CrossRef]
9. Huang, W.; Lin, X. Research progress in laser solid forming of high-performance metallic components at the state key laboratory of solidification processing of china. *3d Print. Addit. Manuf.* **2014**, *1*, 156–165. [CrossRef]
10. Gu, D.; Shi, X.; Poprawe, R.; Bourell, D.L.; Setchi, R.; Zhu, J. Material-structure-performance integrated laser-metal additive manufacturing. *Science* **2021**, *372*, eabg1487. [CrossRef] [PubMed]
11. Herzog, D.; Seyda, V.; Wycisk, E.; Emmelmann, C. Additive manufacturing of metals. *Acta Mater.* **2016**, *117*, 371–392. [CrossRef]
12. Li, Z.; Chen, J.; Sui, S.; Zhong, C.; Lu, X.; Lin, X. The microstructure evolution and tensile properties of inconel 718 fabricated by highdeposition-rate laser directed energy deposition. *Addit. Manuf.* **2020**, *31*, 100941.
13. Huang, B.; Zhai, Y.; Liu, S.; Mao, X. Microstructure anisotropy and its effect on mechanical properties of reduced activation ferritic/martensitic steel fabricated by selective laser melting. *J. Nucl. Mater.* **2018**, *500*, 33–41. [CrossRef]
14. Zhai, Y.; Huang, B.; Mao, X.; Zheng, M. Effect of hot isostatic pressing on microstructure and mechanical properties of CLAM steel produced by selective laser melting. *J. Nucl. Mater.* **2018**, *515*, 111–121. [CrossRef]
15. Boegelein, T.; Dryepondt, S.N.; Pandey, A.; Dawson, K.; Tatlock, G.J. Mechanical response and deformation mechanisms of ferritic oxide dispersion strengthened steel structures produced by selective laser melting. *Acta Mater.* **2015**, *87*, 201–215. [CrossRef]
16. Liao, H.; Wang, X.; Yang, G.; Feng, Y.; Wang, P.; Feng, K. Recent progress of R&D activities on reduced activation ferritic/martensitic steel (CLF-1). *Fusion Eng. Des.* **2019**, *147*, 111235. [CrossRef]
17. Zhong, C.; Chen, J.; Linnenbrink, S.; Gasser, A.; Sui, S.; Poprawe, R. A comparative study of Inconel 718 formed by High Deposition Rate Laser Metal Deposition with GA powder and PREP powder. *Mater. Des.* **2016**, *107*, 386–392. [CrossRef]
18. Lu, J.; Lin, X.; Kang, N.; Huang, W. Characterizations of micro-nano structure and tensile properties of a sc modified almn alloy fabricated by selective laser melting. *Mater. Charact.* **2021**, *178*, 111305. [CrossRef]
19. Song, J.; Deng, Q.; Chen, C.; Hu, D.; Li, Y. Rebuilding of metal components with laser cladding forming. *Appl. Surf. Sci.* **2006**, *252*, 7934–7940. [CrossRef]
20. Huang, Y.; Zeng, X.; Hu, Q.; Zhou, S. Microstructure and interface interaction in laser induction hybrid cladding of Ni-based coating. *Appl. Surf. Sci.* **2009**, *255*, 3940–3945. [CrossRef]
21. Fan, W.; Tan, H.; Lin, X.; Huang, W. Microstructure formation of Ti-6Al-4 V in synchronous induction assisted laser deposition. *Mater. Des.* **2018**, *160*, 1096–1105. [CrossRef]
22. Rosenthal, D. Mathematical Theory of Heat Distribution During Welding and Cutting. *Weld. J.* **1941**, *20*, 220–234.
23. Li, X.P.; Ji, G.; Chen, Z.; Addad, A.; Wu, Y.; Wang, H.W.; Vleugels, J.; Van Humbeeck, J.; Kruth, J.P. Selective laser melting of nano-TiB2 decorated AlSi10Mg alloy with high fracture strength and ductility. *Acta Mater.* **2017**, *129*, 183–193. [CrossRef]
24. Taneike, M.; Sawada, K.; Abe, F. Effect of carbon concentration on precipitation behavior of M23C6 carbides and MX carbonitrides in martensitic 9Cr steel during heat treatment. *Met. Mater. Trans. A* **2004**, *35*, 1255–1262. [CrossRef]
25. Cui, H.; Sun, F.; Chen, K.; Zhang, L.; Wan, R.; Shan, A.; Wu, J. Precipitation behavior of Laves phase in 10%Cr steel X12CrMoWVNbN10-1-1 during short-term creep exposure. *Mater. Sci. Eng. A* **2010**, *527*, 7505–7509. [CrossRef]

metals MDPI

Article

Influence of Relative Humidity and Oxygen Concentration on Corrosion Behaviour of Copper in H_2S-Containing Liquid Petroleum Gas

Xianqiang Li [1], Yuan Lu [2], Qiang Wei [2], Hu Wang [1,*] and Juan Xie [1,*]

1 School of New Energy and Materials, Southwest Petroleum University, Chengdu 610500, China
2 CenerTech Oilfield Chemical Co., Ltd., CNOOC, Tianjin 300450, China
* Correspondence: senty78@hotmail.com (H.W.); jennyx99@126.com (J.X.)

Abstract: In this paper, the influences of relative humidity (*RH*) and concentration of O_2 on copper corrosion in H_2S-containing LPG (liquid petroleum gas) were studied. The corrosion products obtained in different environments were also analysed by scanning electron microscopy (SEM), energy dispersive spectrometry (EDS), grazing incidence X-ray diffraction (GIXRD), X-ray photoelectron spectroscopy (XPS) and Fourier transform infrared spectroscopy (FTIR). In H_2S-containing LPG, *RH* has pronounced influence on the corrosion grade of copper. The variation in the critical point (*CP*) with the *RH* of LPG is a linear relationship. The presence of O_2 in dry H_2S has limited influence on the corrosion of copper. In the presence of different *RH*s, the *CP* always follows a negative exponential function with O_2 concentration. The analysis of different corrosion products implies different corrosion behaviours and mechanisms, which are dependent on the presence or absence of water vapour. The corrosion mechanisms obtained in four different environments were also proposed.

Keywords: copper corrosion; liquefied petroleum gas; H_2S corrosion; SEM; XPS

Citation: Li, X.; Lu, Y.; Wei, Q.; Wang, H.; Xie, J. Influence of Relative Humidity and Oxygen Concentration on Corrosion Behaviour of Copper in H_2S-Containing Liquid Petroleum Gas. *Metals* **2022**, *12*, 2015. https://doi.org/10.3390/met12122015

Academic Editor: David M. Bastidas

Received: 12 October 2022
Accepted: 17 November 2022
Published: 24 November 2022

1. Introduction

Natural gas is widely used in industry and our daily life. As an important existence of natural gas, liquefied petroleum gas (LPG) is more commonly utilized for its accessibility in transportation. In the exploitation and processing of LPG, sulphur removal is inevitable for the sake of alleviating corrosion attack by sulphide existing in produced LPG [1–3]. Various forms of sulphide, such as hydrogen sulphide, mercaptan and carbonyl sulphide, may lead to corrosion of copper components in production and storage facilities. Although the sulphur removal process can eradicate most of the sulphide to some extent, the residual sulphide, in trace amounts, can also be harmful to the corrosion of copper components. Of those common active sulphides, H_2S is the most harmful to copper, even it is presented in low concentration. Although previous studies have revealed the behaviour of H_2S on copper corrosion in LPG, the influence of other environmental factors, such as the presence of O_2 and relative humidity (*RH*), also contribute to corrosion attack.

Recently, the corrosion behaviour of copper in the presence of various sulphides has been extensively studied [4–7]. Echeverria investigated the copper corrosion in SO_2 through atomic force microscopy (AFM). The results showed that the microscopic topography and roughness of the copper surface changed after several weeks in a polluted atmosphere containing SO_2 [8]. Majtás discovered that low concentrations of H_2S can corrode copper parts, resulting in electrical failure of electronic equipment. Additionally, the adsorbed water can promote corrosion attack [9]. Zhu proved that the corrosion rate of copper in an SO_2 environment first increases and then gradually decreases with exposure time. Conversely, the corrosion rate of copper in H_2S increases slowly at first and then sharply declines [10]. Araban studied the corrosion behaviour of copper in different rural atmospheres. The results showed that corrosion product of Cu_2O formed preferentially, in which relative humidity and ammonium sulphate had remarkable influence on the corrosion behaviour [11]. Monzó found that sulphide has an obvious influence on corrosion between the boundary and the centre of a copper sheet. Additionally, elemental sulphur is more corrosive than ethanethiol. The corrosion products of elemental sulphur are in the form of nodule particles, and the ethanethiol is in the form of a uniform film [12]. García found that, at low concentrations of elemental sulphur (5 ppm), mercaptans can significantly promote corrosion. At high concentrations of elemental sulphur (25 ppm), mercaptans inhibit the corrosion of elemental sulphur. Disulphide has an obvious inhibition effect on the corrosion of elemental sulphur [13]. Studies on the corrosion behaviour of copper in the outdoor natural atmosphere have also been reported [14–20]. Kong proved that the uneven corrosion of copper in the atmosphere of Turpan is caused by the dry–wet cycle and the cold–heat cycle [21]. Lopesino believed that the corrosion of copper is more serious when closer to the coast, and the degree of patina coverage depends on the concentration of chloride in the atmosphere [22]. Yan confirmed that the corrosion rate of copper in the atmosphere with sodium chloride is almost 30000 times higher than that in the blank atmosphere [23]. Some other studies focused on the influence of *RH* on the corrosion behaviour of copper [24–26]. Odnevall believed that in the rural atmosphere containing ammonium sulphate, the *RH* of the gas had a great influence on the corrosion behaviour of copper [12]. Sharma proposed that regarding copper in H_2S with low relative humidity, the Cu_2O layer resulted by air has a good protective effect on H_2S. It almost has no protection under high *RH* [27]. Wu proved that the *RH* of the chloride-containing atmosphere is a key factor affecting the corrosion behaviour of copper wires [28]. The corrosion behaviour of chloride on copper has also been extensively studied [29]. Chen proposed that the non-uniform growth of corrosion products on the copper surface in chloride-containing sulphide aqueous solutions resulted in a potential difference between the "thick film" and the "thin film", and this small potential difference accelerated the occurrence of corrosion [30]. Lu believed that chloride ions in the marine atmosphere of Nansha are the key factors to accelerate the corrosion of copper, and the corrosion products are Cu_2O and $Cu_2Cl(OH)_3$ [31]. Schindelholz believed that sodium chloride is favourable for the formation of NaOH-rich diffusion regions, and copper preferentially forms Cu_2O and $Cu(OH)_2$ [32]. There are also few reports on the electrochemical study of corrosion products on copper surfaces [33]. Tran found that the growth of corrosion product films of copper exposed to H_2S-containing subsurface gas has three successive stages: the first stage is a linear growth rate in thin layers (less than 15 nm). In the second stage, the oxidation rate is limited by the diffusion of copper(I) ions through the thicker corrosion layer. The third stage is linear growth [34]. Fiaud believes that both hydrogen sulphide concentration and relative humidity can promote the growth of oxide and sulphide corrosive substances. The growth mechanism of Cu_2O is an electrochemical mechanism, and the growth mechanism of Cu_2S is a mixed chemical and electrochemical mechanism [35]. Some other reports aimed at the corrosion behaviour of copper regarding other aspects, for example, application of theoretical calculations to copper corrosion [5–7,24], the influence of various organic acids on copper corrosion [36,37], the influence of changes in magnetic field on copper corrosion [38] and corrosion behaviour of copper by oxygen plasma [39].

Although some behaviours of copper corrosion in H_2S have been studied, there is still some insufficiency. It is necessary to investigate the corrosion behaviour and mechanism of copper in LPG containing H_2S at different conditions, including the presence of different *RH*s and O_2 concentrations. In this paper, the influence of *RH* and O_2 on the corrosion behaviour of copper in H_2S-containing LPG was studied, the corrosion products on the surface of copper sheets were characterized and analysed and the corresponding corrosion mechanism of H_2S on copper was proposed.

2. Experimental Methods

2.1. Materials

Copper sheets used in corrosion experiments were purchased from Fushun Keruisi Instrument Co., Ltd. Fushun Liaoning Province, China, which strictly follows the requirement of the ASTM standard [40]. The size of cuboid copper sheet is 75 mm × 12.5 mm × 3 mm, with the purity higher than 99.9%. Copper powder (analytical grade, Chengdu Kelon Chemical Co., Ltd. Chengdu, China) was used in X-ray photoelectron spectroscopy (XPS) and FTIR (Fourier transform infrared spectrometry). The purity of powder is 99.5% and the average particle size is 23 μm. The components of LPG are listed in Table 1. H_2S and O_2 gas used in the experiments were purchased from Zhengrong Gas company (Chengdu, China). The purity of H_2S and O_2 is 99.9%.

Table 1. The components of LPG.

Component	Propane	Isobutane	N-Butane
wt./%	70	14	16

2.2. Copper Corrosion Tests

Copper corrosion tests were carried out according to the ASTM standard [40]. The copper sheet was first abrased with 65 μm silicon carbide sandpaper. Then, it was washed with isooctane. The copper surface was polished by 105 μm silicon carbide particles, which were operated with the assistance of isooctane-soaked degreasing cotton. The prepared copper sheets were suspended into the cylinder (the special closed container for corrosion test) in three parallel experiments. Then, high-purity N_2 was used to remove the air inside the cylinder, by ventilating N_2 into the cylinder to substitute the air. Then, to control the mass flow of LPG and H_2S through a flowmeter (FMA5400A, Omega, San Antonio, TX, USA), they were injected into the cylinder and the gas was mixed evenly. Subsequently, the valve was fastened. Finally, the cylinder with mixture of LPG and H_2S was vertically immersed in a water bath at a constant temperature (40 ± 0.5 °C) for 60 ± 5 min. After the experiment time was over, the liquid and gas in the cylinder were discharged. The copper sheets were taken out and compared with the standard colour plate [40]. Next, the corrosion grades of the copper sheets were evaluated. The details of the grade evaluation are shown in Table 2. The standard stipulates that if the corrosion level of a copper sheet reaches 2a and above, it is regarded as unqualified in corrosion.

Table 2. The grading table of copper corrosion standard swatches [40].

Corrosion Level	Copper Colour	Detailed Description
1	mild discolouration	a. pale orange, almost the same as freshly polished copper b. dark orange
2	moderate discolouration	a. fuchsia b. lavender c. multicoloured with lavender blue, silver or both, overlaid on fuchsia d. silver e. brass or golden yellow
3	deep discolouration	a. multicolour magenta overlay brass b. multicolour (malachite green) shown by red and green, no grey
4	corrosion	a. transparent black, dark grey or brown with only malachite green b. graphite or matte black c. glossy black or jet-black glossy black

2.3. Analysis of Corrosion Products

To facilitate the characterization of corrosion products, the concentrations of H_2S and O_2 were increased to 50 ppm and 10 ppm in copper corrosion tests. The copper sheets were used in corrosion tests for characterizations such as scanning electron microscopy (SEM), energy dispersive spectrometry (EDS) and grazing incidence X-ray diffraction (GIXRD). Samples of corrosion products for XPS and FTIR were prepared with powder, which could provide better results than copper sheets. The copper powder samples were applied by placing 0.1 g of copper powder in a glass sample bottle. Then, the glass sample bottle was suspended in a cylinder to conduct a copper corrosion experiment.

SEM (Model EVO MA15, ZEISS, Jena, Germany) was used to observe the morphology of the corrosion products on the copper surface. EDS (Model X-MaxN, OXFORD INSTRUMENTS Company, Abingdon, UK) was used to analyse the elemental composition of the corrosion products on the copper sheet surface. XPS (Nexsa type, Thermo Scientific, Waltham, MA, USA) was used to analyse the elemental composition and valence distribution of the corrosion products on the copper powder surface. The original XPS image was fitted with Casa XPS software. GIXRD (SmartLab 9 kw, Rigaku, Tokyo, Japan) was used to analyse the phase composition of corrosion products on the copper sheet surface. FTIR (INVENIO R, Bruker Optik GmbH, Bremen, Germany) was carried out to test the infrared spectrum of the corrosion products on the copper powder surface.

3. Results and Discussion

3.1. *Influence of Humidity on Copper Corrosion in H_2S-Containing LPG*

The corrosion behaviour of copper in H_2S-containing LPG is very sensitive to the gas humidity. The higher the gas humidity, the more easily copper is corroded by H_2S. Figure 1a shows the variation in the corrosion grade of copper in H_2S-containing LPG at different gas humidities. In LPG containing 3 ppm H_2S, the corrosion grade of copper gradually intensifies with the increase in gas humidity. At 0–30% *RH*, there is no significant corrosion on the copper surface (corrosion grade is 1a). At 50–100% *RH*, the surface of the copper sheet begins to corrode (at 50% *RH*, the copper corrosion grade is 2a). The degree of corrosion varies with humidity. The copper corrosion grade reaches 2e at 100% *RH*. Meanwhile, the variation in the copper corrosion grade in LPG without H_2S in the presence of different gas humidities is used as comparison, in which the copper surface does not corrode at all humidities (corrosion grade is 1a).

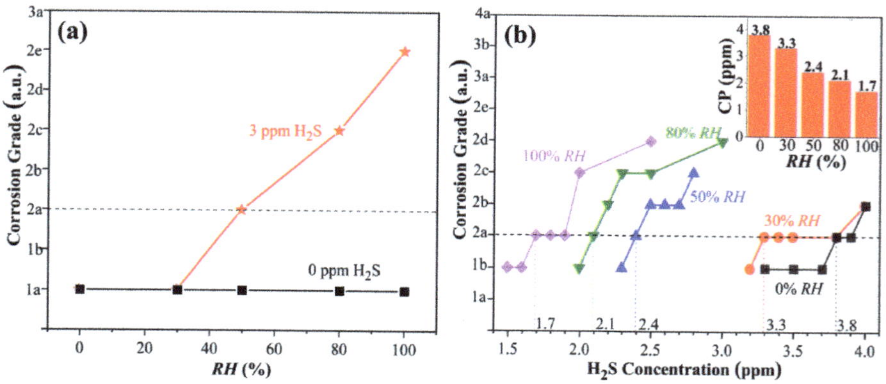

Figure 1. (**a**) Variation in copper corrosion grade in H_2S-containing LPG with humidity; (**b**) the influence of H_2S concentration and *RH* on the corrosion grade of copper in LPG.

The critical point (*CP*) of copper corrosion in H_2S-containing LPG decreases with the increase in the relative humidity of LPG. The *CP* is defined as the lowest H_2S concentration to reach the corrosion grade of 2a at specific environmental conditions. Figure 1b shows the influence of H_2S concentration and *RH* on the corrosion grade of copper in LPG. It can be seen that the *CP* gradually decreases with the increase in gas humidity. From 0% *RH* to 100% *RH*, the *CP* decreases from 3.8 ppm H_2S to 1.7 ppm H_2S. It indicates that higher humidity is more beneficial to the corrosion process. In higher humidity, the thin film of water at the copper surface forms more easily, which provides an electrolyte environment for H_2S dissolution and electrochemical corrosion. With the increase in *RH*, the thickness of the water film at the interface increases, which will provide a better condition for electrochemical corrosion. Therefore, the corrosion attack is more severe at higher *RH* and the *CP* would be lowered with the increase in *RH*.

The influence of *RH* on the *CP* is shown in Figure 2. The results of the *CP* at different *RH*s show a linear relationship. The fitted data obey Equation (1) as follows:

$$CP = -0.021\ RH + 3.77\ ,\ R_0 = 0.96 \qquad (1)$$

where *CP* is the critical point of copper corrosion, corresponding to the lowest H_2S concentration for reaching the corrosion grade of 2a. *RH* is the relative humidity for the corrosion test. *R* is the coefficient of determination. The results imply that with the increase in *RH*, the thickness of the water film formed at the copper surface increases accordingly. The thicker water film is more favourable for H_2S dissolution. Consequently, the electrochemical corrosion process is enhanced.

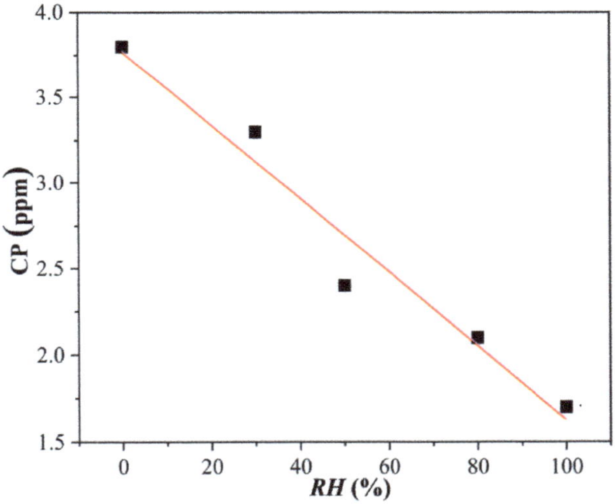

Figure 2. The influence of *RH* on *CP* of copper corrosion in H_2S-containing LPG.

3.2. Synergistic Effect of Oxygen and Humidity on Copper Corrosion in H_2S-Containing LPG

A small amount of O_2 has limited influence on the corrosion of copper in dry LPG (0% *RH*). Figure 3a shows the variation in copper corrosion grade with O_2 content in LPG (0% *RH*) containing H_2S. As can be seen from Figure 3a, in pure LPG (without H_2S), the corrosion grade of copper does not change with the increase in O_2 content (corrosion grade is 1a). No apparent corrosion happened at the copper surface at such condition. In LPG containing trace H_2S (1 ppm), slight corrosion on the copper surface appears with the increase in O_2 content. Among them, there is no apparent corrosion on the copper surface from 0 to 5 ppm O_2, and the corrosion grade of copper sheets in 10 ppm O_2 begins to rise to 1b, which is also below 2a.

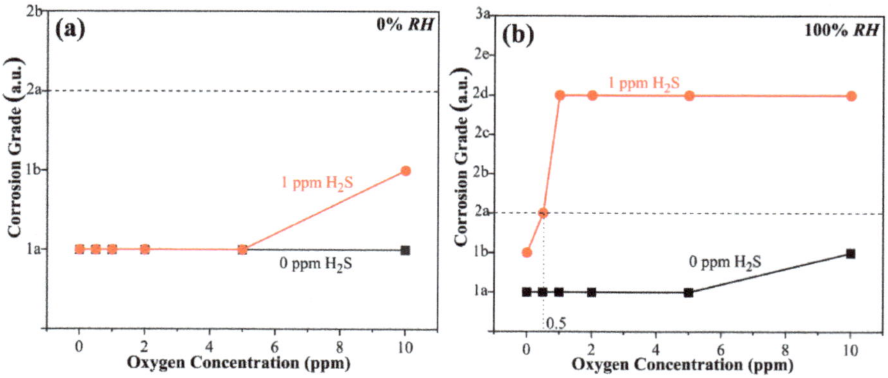

Figure 3. Influence of oxygen concentration on copper corrosion grade in H_2S-containing LPG at (**a**) 0% *RH*, (**b**) 100% *RH*.

A small amount of O_2 has a pronounced effect on the corrosion of copper in wet LPG (100% *RH*). Figure 3b is the variation in copper corrosion grade in H_2S-containing LPG (100% *RH*) with O_2 concentration. It demonstrates that in the absence of H_2S, the copper corrosion grade can hardly be changed with the increase in O_2 content. The copper surface does not corrode in 0–5 ppm O_2, displaying a corrosion grade of 1a. In the presence of 10 ppm O_2, the copper corrosion grade is slightly promoted to 1b. In the presence of a trace amount of H_2S (1 ppm) at 100% *RH*, the degree of corrosion is sharply intensified with the increase in O_2 content. An amount of 0.5 ppm of O_2 can lead to unqualified copper corrosion (grade 2a). When the O_2 content increases to 1 ppm, the corrosion grade rapidly climbs to grade 2d. When the content of O_2 continues to increase, the corrosion grade is stabilized at grade 2d.

Figure 3 reveals that the coexistence of gas humidity and O_2 has a notable synergistic effect on the corrosion of copper in LPG in the presence of H_2S. Compared with pure LPG (0% *RH*, 0 ppm O_2), the copper in LPG containing wet H_2S and O_2 is more easily corroded.

In order to further study the synergistic effect of oxygen and humidity on the *CP* of copper corrosion in H_2S-containing LPG, the corrosion behaviour of copper in H_2S-containing LPG in the presence of different oxygen concentrations was studied at 0% *RH*, 30% *RH*, 50% *RH*, 80% *RH* and 100% *RH*, respectively. The results are shown in Figure 4. It can be seen that at every *RH* condition, the *CP* gradually decreases with the increase in O_2 concentration. At the same oxygen concentration, *CP* gradually declines with the increase in the gas humidity.

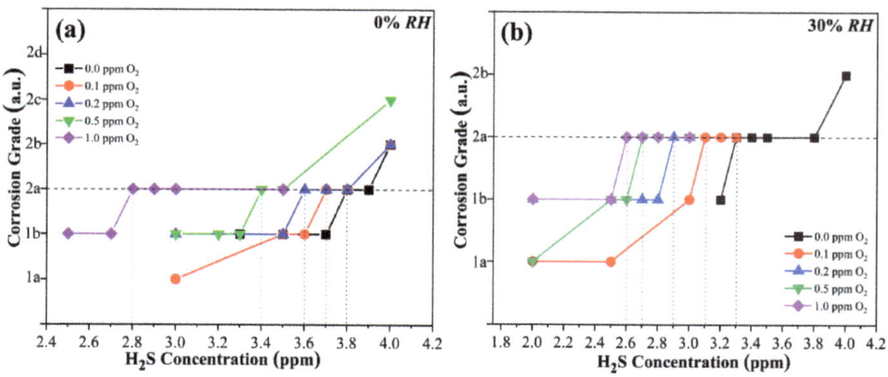

Figure 4. *Cont.*

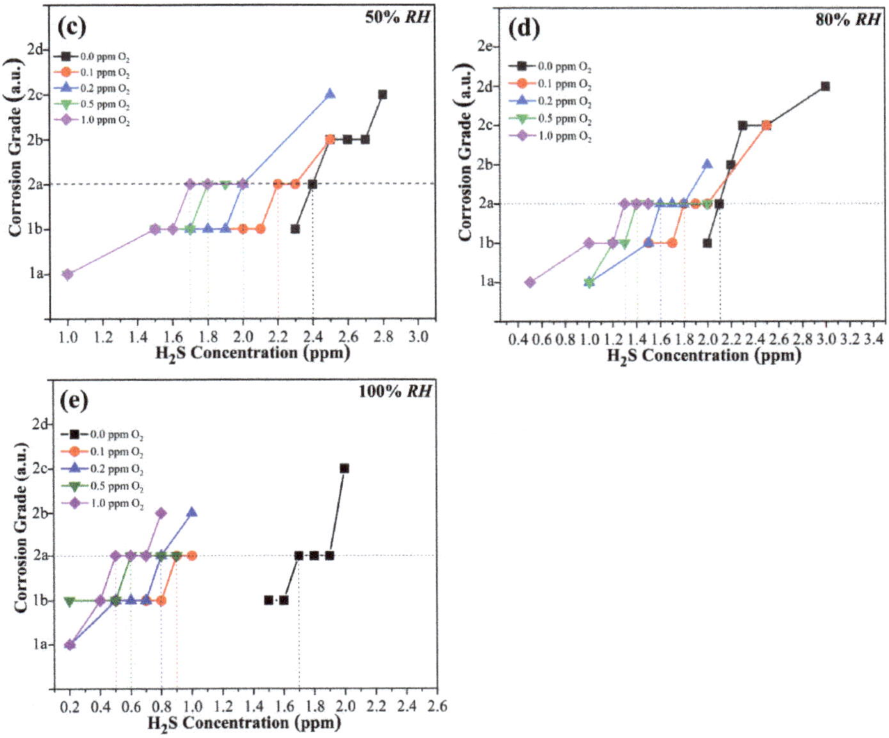

Figure 4. Influence of oxygen content on *CP* at (**a**) 0% *RH*, (**b**) 30% *RH*, (**c**) 50% *RH*, (**d**) 80% *RH*, (**e**) 100% *RH*.

More precise behaviour can be illustrated by interpreting the relationship between the *CP* and O_2 concentration at different *RH*s, as is shown in Figure 5. The discussion is carried out at different *RH*s.

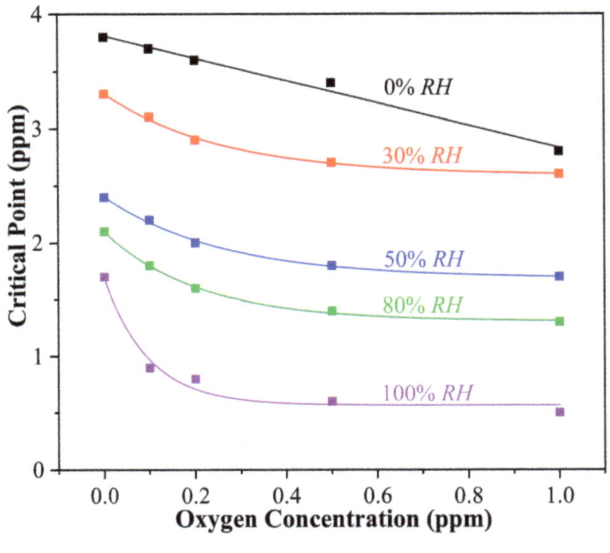

Figure 5. Fitted data of the *CP* in different LPG environments.

(a) In the absence of water (0% *RH*), *CP* follows a linear relationship with O_2 concentration. The fitted data obey Equation (2) as follows:

$$CP_{0\%} = 3.81 - 0.978\, C_O\,,\ \ R_1 = 0.98 \tag{2}$$

where $CP_{x\%}$ is the critical point of copper corrosion in H_2S-containing LPG (x% *RH*), corresponding to the minimum H_2S concentration for corrosion grade 2a. C_O is the oxygen concentration of LPG in the copper corrosion test. *R* is the coefficient of determination.

(b) In the presence of water (30% *RH*), *CP* follows a negative exponential function with the O_2 concentration. The fitted data obey Equation (3) as follows:

$$CP_{30\%} = 0.720\, e^{-3.82\, C_O} + 2.585\ ,\ \ R_2 = 0.99 \tag{3}$$

(c) In the presence of water (50% *RH*), *CP* follows a negative exponential function with the O_2 concentration. The fitted data obey Equation (4) as follows:

$$CP_{50\%} = 0.720\, e^{-3.82\, C_O} + 1.685\ ,\ \ R_3 = 0.99 \tag{4}$$

(d) In the presence of water (80% *RH*), *CP* follows a negative exponential function with the O_2 concentration. The fitted data obey Equation (5) as follows:

$$CP_{80\%} = 0.793\, e^{-4.74\, C_O} + 1.305\ ,\ \ R_4 = 0.99 \tag{5}$$

(e) In the presence of water (100% *RH*), *CP* follows a negative exponential function with the O_2 concentration. The fitted data obey Equation (6) as follows:

$$CP_{100\%} = 1.123\, e^{-10.20\, C_O} + 0.564\ \ ,\ \ R_5 = 0.98 \tag{6}$$

In the absence of H_2O, the contribution of O_2 to copper corrosion is relatively even, which is consistent with a previous report [27]. However, the presence of H_2O makes copper corrosion more sensitive to O_2 even at low O_2 concentration. The formation of a water film on the copper surface makes the corrosion process different. According to the Arrhenius Equation, Equations (7) and (8), in kinetics, when the temperature of the reaction system is constant, the rate constant of a specific chemical reaction is related to the activation energy of the reaction. The lower the activation energy of the reaction, the faster the reaction rate is. In the presence of H_2O, the activation energy in the reaction system decreases (Equation (9)). Compared with the reaction system without H_2O, the reaction rate constant (*k*) is larger, so the reaction rate is faster. This explains why the *CP* at 0% *RH* is higher than the *CP* in the presence of H_2O.

$$k_{(0)} = A\, e^{(-E_{a(0)}/RT)} \tag{7}$$

$$k_{(c)} = A\, e^{(-E_{a(c)}/RT)} \tag{8}$$

from Equations (7) and (8):

$$E_{a(c)} = E_{a(0)} - RT \ln\left(k_{(c)}/k_{(0)}\right) \tag{9}$$

where $k_{(0)}$ is the rate constant of the reaction, $k_{(c)}$ is the rate constant of the reaction after adding the catalyst, $E_{a(0)}$ is the activation energy of the reaction ($kJ \cdot mol^{-1}$), $E_{a(c)}$ is the activation energy of the reaction after adding the catalyst ($kJ \cdot mol^{-1}$), *A* is the pre-exponential factor, *e* is the natural base (2.718), *R* is the gas constant (8.314 $J \cdot mol^{-1} \cdot K^{-1}$) and *T* is the thermodynamic temperature (*K*). The presence of the water film, which acts as a catalyst in the system at the interface, changes the kinetics of the corrosion process.

The variation in the *CP* with O_2 at different *RHs* can also be explained by Equations (7)–(9). It is well known that the presence of a water film at the interface can reduce the activation of the reaction, which can increase the number of activated molecules in the reaction system by increasing the number of effective collisions. Thus, it significantly accelerates the reaction rate. A higher *RH* in the reaction system means a thicker water vapour film on the copper surface, which implies greater effectiveness in promoting the corrosion process. When the reaction concentration is constant, the thicker water film can generate more activated molecules in the reaction system, the reaction rate constant (k) is larger and more effective collisions are generated per unit time to form more Cu_2S and Cu_2O.

3.3. *Corrosion Mechanism of Copper in Different H_2S-Containing LPG Environments*

3.3.1. Surface Morphologies after Corrosion

In different environments, LPG with H_2S, $H_2S + H_2O$, $H_2S + O_2$ and $H_2S + O_2 + H_2O$, the microscopic morphologies of the corrosion products on the copper surface sheet are shown in Figure 6. In the absence of H_2S, the copper surface displays an uncorroded appearance with grooves of abrasion. When corroded in H_2S-containing LPG, the copper surface is evenly covered with a thick corrosion product film. The corrosion products are in the shape of a regular hexagon with sharp edges and corners. In LPG (100% *RH*) containing $H_2S + H_2O$, the copper surface is evenly covered with a thick layer of corrosion product film. The corrosion products are spherical and accumulate at the grooves of scratches, indicating that the nucleation and growth of corrosion tend to preferentially happen at grooves of scratches [41]. A similar phenomenon also appears in other environments. In $H_2S + O_2$, the amount of corrosion products is significantly reduced. The corrosion products are sporadically distributed on the copper surface. The white particles of corrosion products are irregular in shape and size. It can be seen from the morphology that the general corrosion at this condition is significantly reduced, which is consistent with the previous experimental results. The corrosion attack happens at localized active sites, not on the whole surface. In $H_2S + H_2O$ (100% *RH*) + O_2, it exhibits a thick corrosion product film on the copper surface. Some irregular white corrosion products attach on the film surface.

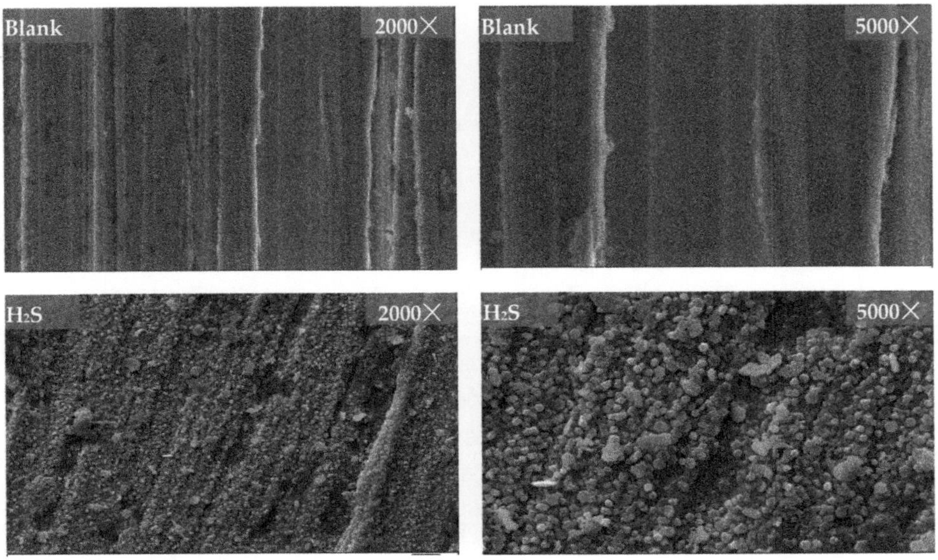

Figure 6. *Cont.*

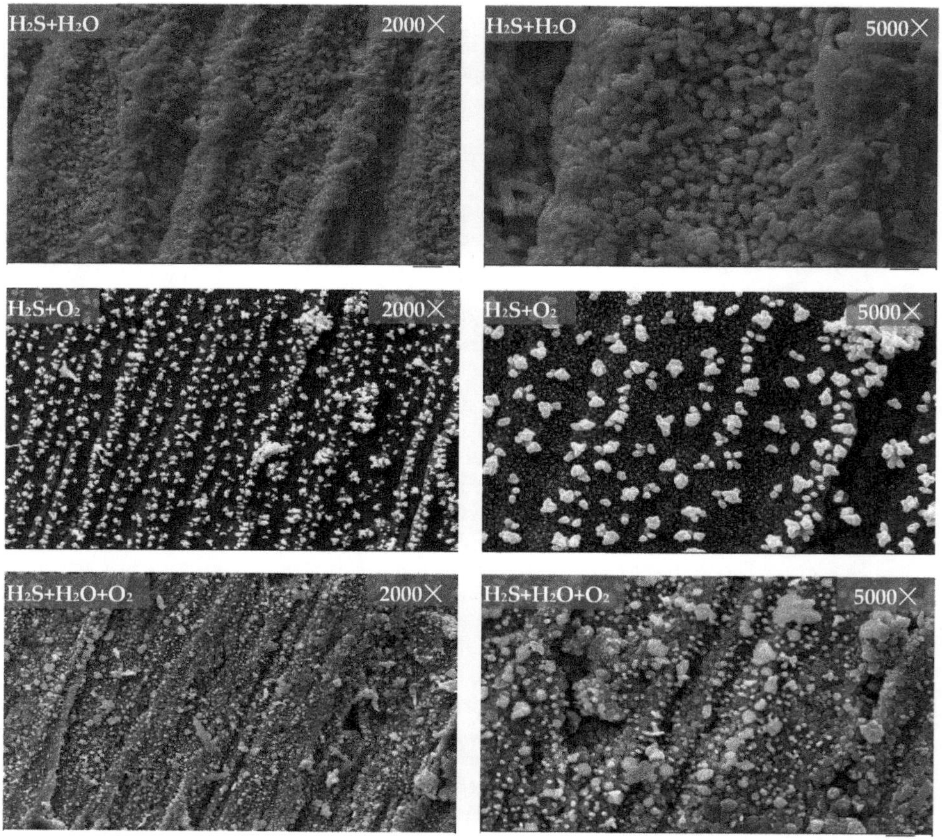

Figure 6. SEM images of corrosion products of copper sheets after corrosion tests in H_2S, $H_2S + H_2O$, $H_2S + O_2$ and $H_2S + O_2 + H_2O$.

3.3.2. EDS Analysis of Corrosion Products

EDS was used to analyse the elemental information of corrosion products. Figure 7 and Table 3 manifest the elemental content of copper corrosion products at four medium conditions (H_2S, $H_2S + H_2O$, $H_2S + O_2$ and $H_2S + H_2O + O_2$). It can be seen from the results that the corrosion products of LPG in H_2S mainly contain S and Cu, indicating that the corrosion products are only composed of copper sulphides. The corrosion products of LPG containing $H_2S + H_2O$ mainly contain S, Cu and O, implying that they are mainly composed of copper sulphides and oxides. It is also possible that oxides were generated by exposure of the sample to air. The corrosion products in H_2S+O_2 mainly contain S, Cu and O. It shows that the content of S is much higher than that of O, indicating that the corrosion products are mainly composed of a large amount of copper sulphides and a small amount of copper oxides. The corrosion products in $H_2S + H_2O + O_2$ mainly contain S, Cu and O elements, meaning that the corrosion products are mainly composed of copper oxides and sulphides. The content of O is much higher than that of S, implying that the existence of H_2O and O_2 is more favourable for the growth of copper oxide. In the presence of H_2O (100% *RH*), it is more favourable to form a water film on the copper surface, which in turn leads to the dissolution and diffusion of oxygen. The electrochemical corrosion happens with the cathode process of oxygen depolarization reaction. Additionally, the dissolution of H_2S in water film leads to the emergence of H^+. The hydrogen depolarization reaction as a cathodic process also happens. The two cathodic processes occur at the same time in $H_2S + O_2 + H_2O$, which induces the synergistic corrosion effects.

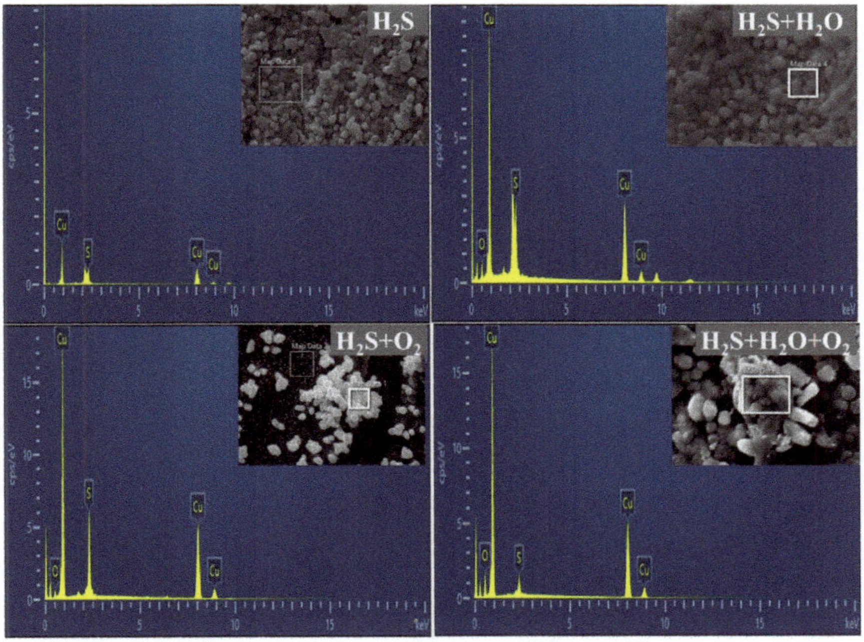

Figure 7. EDS images of corrosion products in different LPG environments.

Table 3. The elemental content of corrosion products at four different corrosion conditions.

Element	H_2S		$H_2S + H_2O$		$H_2S + O_2$		$H_2S + O_2 + H_2O$	
	wt.%	Atomic%	wt.%	Atomic%	wt.%	Atomic%	wt.%	Atomic%
O	-	-	5.18	16.61	2.85	9.20	8.40	25.99
S	9.94	17.95	8.66	13.85	14.98	24.10	3.50	5.40
Cu	90.06	82.05	86.15	69.53	82.17	66.70	88.09	68.60
Total	100.00	100.00	100.00	100.00	100.00	100.00	100.00	100.00

3.3.3. GIXRD Analysis of Corrosion Products

In order to further reveal the corrosion mechanism of copper sheets in different LPG environments, GIXRD was used to analyse the corroded copper sheets in the presence of H_2S, $H_2S + H_2O$, $H_2S + O_2$ and $H_2S + H_2O + O_2$, respectively, and the incident angle of GIXRD was 0.7°. The results in Figure 8 show that the spectrum of corroded copper in H_2S-containing LPG is mainly the diffraction peaks of Cu and Cu_2S, in which Cu_2S preferentially grows on the (−536) crystal plane. The spectrum of corroded copper in LPG containing H_2S+H_2O is mainly the diffraction peaks of Cu and Cu_2S, in which Cu_2S preferentially grows on the (−232) crystal plane. The expected spectrum of Cu_2O cannot be obtained, which has been proven by EDS. This is due to a too little amount of corrosion products. The spectrum of corrosion products in LPG containing H_2S+O_2 is similar to the diffraction peaks of corrosion products at H_2S conditions, mainly Cu and Cu_2S diffraction peaks. Similarly, Cu_2S grows preferentially on the (−536) crystal plane. The spectrum of corrosion products in LPG containing H_2S+O_2+H_2O is mainly Cu, Cu_2S and Cu_2O. The diffraction peak of Cu_2O is obviously stronger than that of Cu_2S, indicating that the conditions are more preferable to the growth of Cu_2O [34,35,42]. In addition, in this circumstance Cu_2S preferentially grows along the (034) crystal plane, which is different from other environmental conditions.

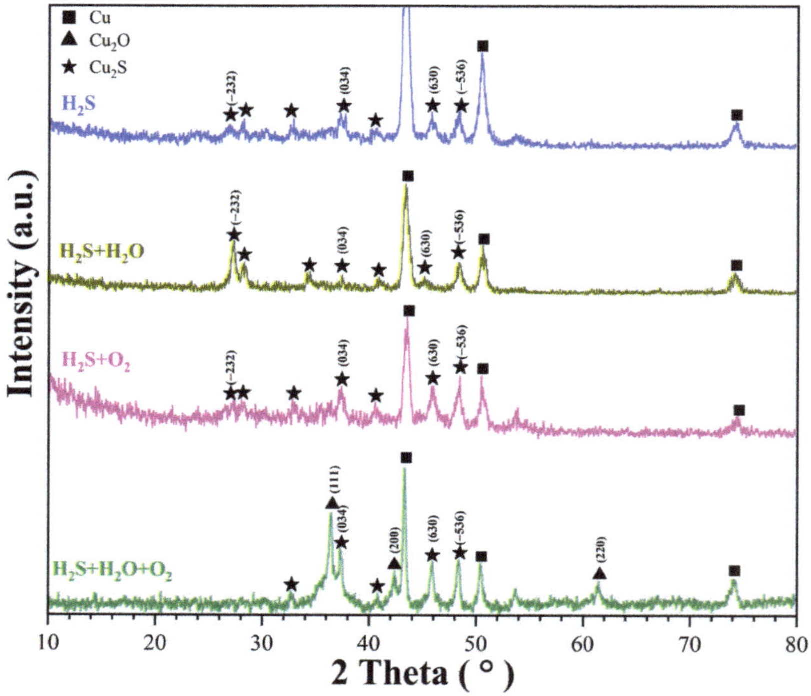

Figure 8. GIXRD patterns of corrosion products at different conditions.

3.3.4. XPS Analysis of Corrosion Products

XPS was also applied to analyse the valence states of corrosion products of copper sheets in different LPG environments, including H_2S, $H_2S + H_2O$, $H_2S + O_2$ and $H_2S + H_2O + O_2$. The results are shown in Figure 9. Figure 9a is a comparison diagram of the Cu 2p spectrum of the corrosion products in different LPG environments. In the Cu 2p spectrum of the corrosion products in H_2S, the peaks at 932.75 eV and 945 eV correspond to the characteristic peak of Cu_2S and the satellite peak of Cu^+ at the 2p3/2 orbital, respectively. The peaks at 932.77 eV, 934.10 eV and 943.00 eV in Cu 2p spectrum of corrosion products in H_2S+H_2O correspond to the characteristic peaks of Cu_2S, CuO and the satellite peaks of Cu^{2+} at the 2p3/2 orbital, respectively. The peaks at 932.70 eV, 934.27 eV and 943.00 eV in the Cu 2p spectrum of corrosion products in H_2S+O_2 correspond to the characteristic peaks of Cu_2S, CuO and the satellite of Cu^{2+} at the 2p3/2 orbit, respectively. The peaks at 932.80 eV, 932.51 eV and 945.00 eV in the Cu 2p spectrum of the corrosion products with $H_2S + H_2O + O_2$ correspond to the characteristic peaks of Cu_2S, Cu_2O and the satellite of Cu^+ at the 2p3/2 orbital, respectively, which are consistent with the XRD results. Figure 9b is the analysis result of the high-resolution S 2p spectrum. As seen in Figure 9b, in different LPG environments, the XPS signal of S is weak and exists in the form of Cu_2S.

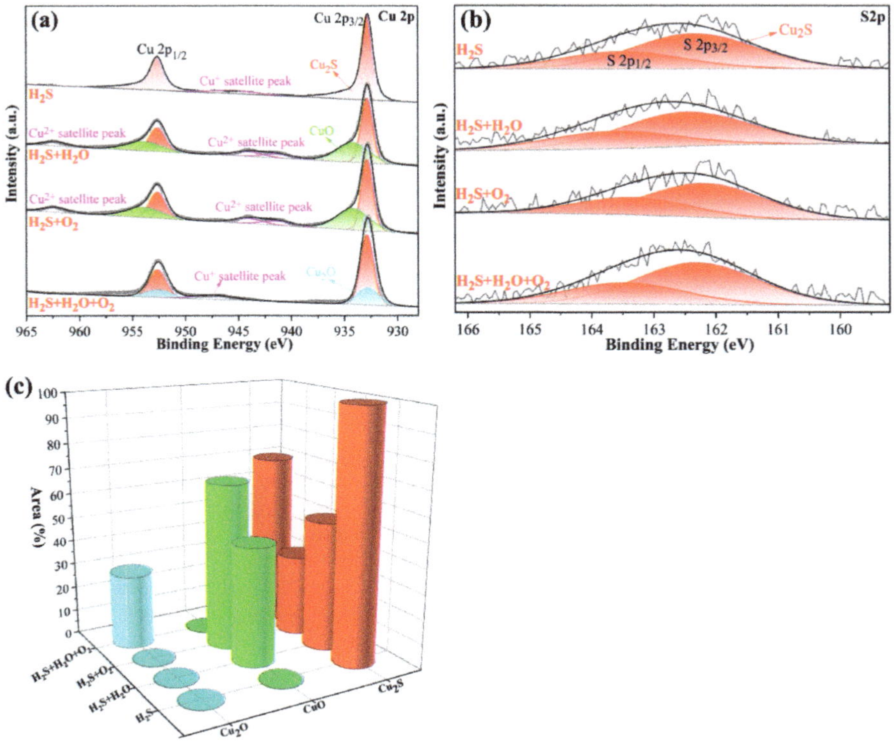

Figure 9. XPS images of corrosion products in different LPG environments: (**a**) Cu 2p data and fits; (**b**) S 2p data and fits; (**c**) the content ratio of each phase in Cu 2p spectrum.

Figure 9c shows the comparative analysis results of the content of Cu_2S, Cu_2O and CuO in the Cu 2p spectrum in different LPG environments. The detailed information of each phase is shown in Table 4. It can be seen from Figure 9c that Cu_2S exists in all four conditions, and the proportion is the highest in pure H_2S. With the addition of H_2O and O_2, the content of Cu_2S decreases and the content of CuO increases gradually. In the presence of H_2O + O_2, Cu_2S content begins to rise again, and Cu_2O appears in large quantities, which agrees with the previous EDS and XRD results.

Table 4. The content ratio of each phase in the corrosion products in the Cu 2p spectrum.

Element	Peak Type	Condition	BE (ev)	Area%
Cu	Cu_2S	H_2S	932.75	100.00
		$H_2S + H_2O$	932.77	52.74
		$H_2S + O_2$	932.70	32.56
		$H_2S + H_2O + O_2$	932.80	70.26
	Cu_2O	H_2S	-	-
		$H_2S + H_2O$	-	-
		$H_2S + O_2$	-	-
		$H_2S + H_2O + O_2$	932.51	29.74
	CuO	H_2S	-	-
		$H_2S + H_2O$	934.10	47.26
		$H_2S + O_2$	934.27	67.44
		$H_2S + H_2O + O_2$	-	-

3.3.5. FTIR Analysis of Corrosion Products

FTIR was used to analyse the corrosion products in different LPG environments. The results are shown in Figure 10. It compares the FTIR spectra of copper powder and copper powder with four different corrosion products. The full FTIR spectra of all corrosion products are similar to the blank. The absorption peak at 3743 cm^{-1} represents the stretching vibration absorption peak of free-state O-H in the impurities on the copper powder surface. Compared with the blank, the weakening of the peak intensity is caused by the accumulation of a thick layer of corrosion products on the copper powder surface. Compared with blank, the absorption peak of 3446 cm^{-1} is a composite peak, which is composed of a stretching vibration absorption peak of the associative O-H and Cu(I)-S on the copper powder surface. The phenomenon of increased peak intensity is due to the presence of a large amount of Cu_2S (3470 cm^{-1}) after corrosion. The absorption peaks of 2923 cm^{-1} and 2854 cm^{-1} represent the stretching vibration absorption peaks of the C-H bond in the methylene group. The phenomenon of peak intensity enhancement is caused by the residual LPG in the corrosion products. In the FTIR spectrum of the bank, there are many unknown absorption peaks around 1500 cm^{-1}, which should be caused by impurities of the copper powder surface. Figure 10b is an FTIR comparison of 920–1230 cm^{-1}. Compared with the blank, the absorption peak at 1160 cm^{-1} is significantly enhanced, which can be attributed to the stretching vibration of Cu(I)-S (1145 cm^{-1}). The reason why the peak intensity at this position is similar to the blank can be ascribed to little corrosion products on the copper powder surface in $H_2S + O_2$. Compared with the blank, the peak position of 1081 cm^{-1} appears blue-shifted, which is caused by the vibration of a large amount of Cu(I)-S (1115 cm^{-1}). This phenomenon is more pronounced in conditions with more corrosion products. Figure 10c shows the FTIR comparison between 400 cm^{-1} and 800 cm^{-1} for the four corrosion products and the blank. Compared with the blank, the apparent enhancement of the absorption peak at 713 cm^{-1} is due to the in-plane rocking vibration of C-H in the methylene group of the residual LPG adsorbed in the corrosion product. As for the weak absorption peak at 619 cm^{-1} in H_2S and $H_2S + H_2O$, it can be regarded as the stretching vibration peak of Cu(I)-S (618 cm^{-1}). In $H_2S + O_2$ and $H_2S + H_2O + O_2$, a strong absorption characteristic peak at 619 cm^{-1} can be regarded as the composite absorption peak of Cu_2S and Cu_2O (the absorption peak positions of Cu(I)-S (618 cm^{-1}) and Cu(I)-O (614 cm^{-1}) are very similar). The strong absorption peak at 499 cm^{-1} is due to the stretching vibration of Cu=O [42–45].

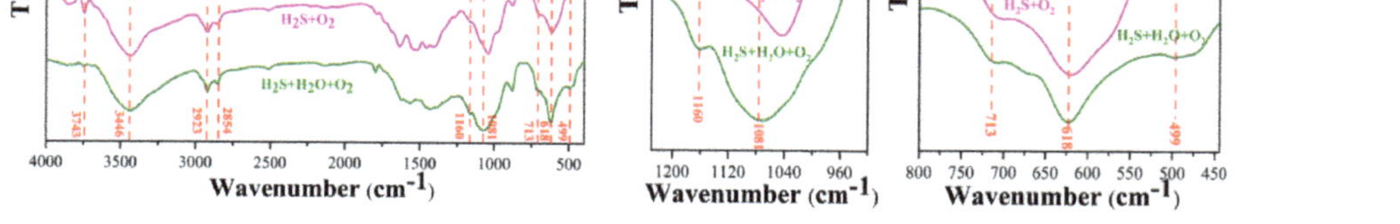

Figure 10. FTIR images of corrosion products on copper powder in different LPG environments: (**a**) FTIR full spectrum, (**b**) FTIR images of 920~1230 cm^{-1}, (**c**) FTIR images of 445~800 cm^{-1}.

The results of the analysis on the differences in the absorption peaks in the FTIR reveal that Cu_2S will be formed on the copper surface in the four conditions. The presence of H_2O promotes the production of CuO.

3.4. Corrosion Mechanism of Copper in Different LPG Environments

Based on the results of copper corrosion tests and corrosion product analysis, the copper corrosion mechanisms of copper at different LPG environments can be proposed as follows.

3.4.1. Corrosion Mechanism in H_2S

The corrosion mechanism of copper in LPG only containing H_2S is a chemical corrosion process. The corrosion process is carried out according to the reaction Equations (10) and (11). Figure 11 is a schematic diagram of the corrosion steps in H_2S-containing LPG. In the first step, H_2S gas in the LPG is adsorbed on the copper surface. Then, H_2S reacts with Cu atoms at the surface to generate H_2 and Cu_2S. The whole corrosion process is a chemical process. Electrochemical corrosion cannot occur due to the absence of the electrolyte.

$$H_2S_{(g)} \rightleftharpoons H_2S_{(ads)} \tag{10}$$

$$H_2S_{(ads)} + 2Cu \rightleftharpoons Cu_2S + H_2 \tag{11}$$

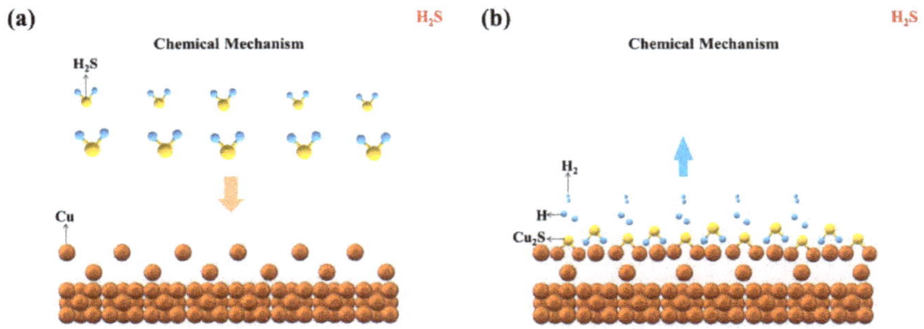

Figure 11. Schematic diagram of corrosion steps of copper sheet in H_2S-containing LPG. (**a**) step1. (**b**) step2.

3.4.2. Corrosion Mechanism in H_2S+H_2O

The corrosion mechanism of copper in LPG containing $H_2S + H_2O$ is an electrochemical corrosion process. The corrosion process is carried out according to the reaction Equations (12)–(16). Figure 12 shows the schematic corrosion steps in LPG with H_2S + H_2O. The process of electrochemical corrosion can be explained by Figure 12a–c. Firstly, H_2O exists in the form of a water film on the copper surface. H_2S in LPG dissolves into the water film and further hydrolyses to form a great amount of HS^-, H_3O^- (hydronium ions) and a small amount of S^{2-}. Furthermore, the process of the anodic reaction is losing electrons of a Cu atom to form Cu_2S with S^{2-}, as expressed in Equation (15). The cathodic reaction can be undertaken as Equation (16), which is the traditional hydrogen depolarization reaction. At this time, the massive consumption of S^{2-} further promotes the dissolution and hydrolysis of H_2S in LPG, which continuously accelerates the corrosion of copper and results in a great amount of Cu_2S precipitation. However, when the water film is insufficient (low humidity), it is difficult to form an effective electrochemical corrosion environment. In this circumstance, the chemical process may be the main reason for corrosion. This is consistent with the results of XRD, XPS and FTIR. A small amount of CuO in XPS and FTIR is likely due to the contamination of oxidization in air.

$$H_2S_{(g)} \rightleftharpoons H_2S_{(l)} \tag{12}$$

$$H_2S_{(l)} + H_2O \rightleftharpoons HS^- + H_3O^+ \tag{13}$$

$$HS^- + H_2O \rightleftharpoons S^{2-} + H_3O^+ \tag{14}$$

$$2Cu + S^{2-} \rightleftharpoons Cu_2S + 2e^- \quad (15)$$

$$2H_3O^+ + 2e^- \rightleftharpoons 2H_2O + H_2 \quad (16)$$

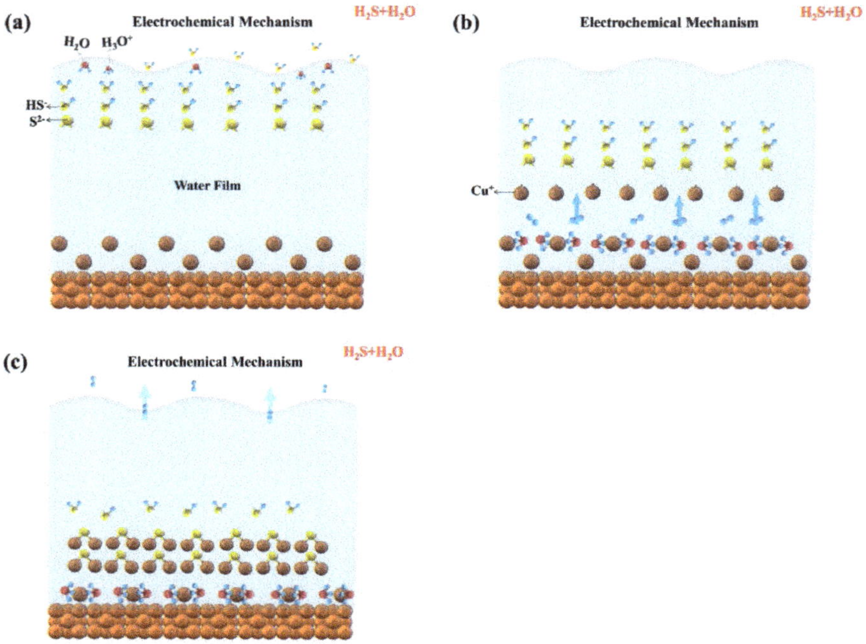

Figure 12. Schematic diagram of corrosion steps of copper sheet in LPG with H_2S+H_2O. (**a**) step1. (**b**) step2. (**c**) step3.

3.4.3. Corrosion Mechanism in H_2S + O_2

The corrosion mechanism of copper in LPG with H_2S + O_2 is a chemical corrosion process. The corrosion process is carried out according to Equations (10), (11) and (17)–(19). Figure 13 displays a schematic diagram of the corrosion process in LPG containing H_2S + O_2. In the first step, H_2S and O_2 molecules in LPG adsorb on the copper surface. Then, the chemical reaction of copper with H_2S and O_2 molecules generate Cu_2S, H_2 and Cu_2O, respectively. Part of Cu_2O is further oxidized to CuO by O_2. The corrosion mechanism shows that the presence of only O_2 will not significantly promote the corrosion grade of the copper sheet, which is consistent with the experimental results. In this condition, the peak of CuO does not appear in the XRD results. However, a small amount of CuO was detected in XPS, indicating that the content of CuO is quite low in corrosion products. With the FTIR spectrum, the amount of Cu_2O is significantly more than CuO, which implies Cu_2O is more stable than CuO.

$$O_{2(g)} \rightleftharpoons O_{2(ads)} \quad (17)$$

$$O_{2(ads)} + 4Cu \rightleftharpoons 2Cu_2O \quad (18)$$

$$O_{2(ads)} + 2Cu_2O \rightleftharpoons 4CuO \quad (19)$$

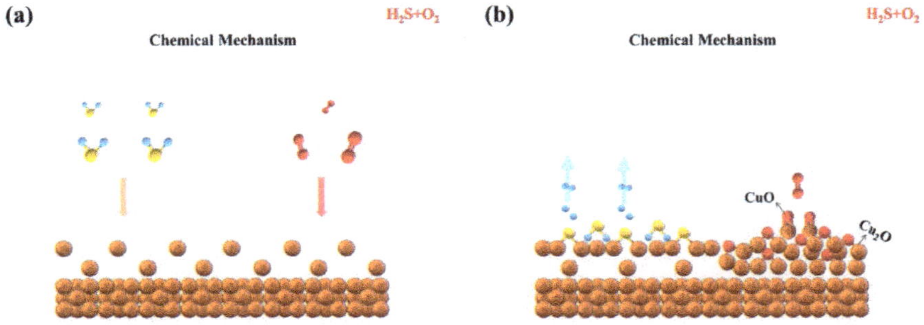

Figure 13. Schematic diagram of corrosion steps of copper sheet in LPG with $H_2S + O_2$. (**a**) step1. (**b**) step2.

3.4.4. Corrosion Mechanism in $H_2S + H_2O + O_2$

The corrosion mechanism of copper in LPG containing $H_2S + H_2O + O_2$ is an electrochemical corrosion process. The corrosion process is carried out according to the Equations (12)–(16) and (20)–(22). The electrochemical corrosion process can be illustrated by the schematic diagram in Figure 14. Firstly, H_2O in LPG exists in the form of a water film on the copper surface. Two electrochemical reactions happen at the interface. The dissolution of H_2S in the water film leads to the electrochemical reaction in Equations (12)–(16). Meanwhile, the dissolution and diffusion of O_2 also cause another electrochemical corrosion reaction, as in Equations (20)–(22), which is the oxygen depolarization process. After H_2S and O_2 molecules in LPG dissolve into the water film to form $H_2S_{(l)}$ and $O_{2(l)}$, $H_2S_{(l)}$ is further hydrolysed into a large amount of HS^-, H_3O^+ and a small amount of S^{2-}. Cu loses electrons to form Cu_2S and Cu_2O with S^{2-} and OH^-, according to the anodic reaction of Equations (15) and (22) [46]. H_3O^+ and $O_{2(l)}$ obtain electrons from copper to form H_2, H_2O and OH^-, which are regarded as cathode reactions (Equations (16) and (21)) [34,35,47]. Because the H_3O^+ in the water is consumed in large quantities, it is conducive to the ionization of water to move to the right to generate a large amount of OH^-, which increases the pH of the water film. A large amount of OH^- is not only conducive to the dissolution of H_2S in LPG in the water film, improving the solubility of H_2S, but is also conducive to the continuous hydrolysis of $H_2S_{(l)}$ and HS^- in the water film to form a large amount of HS^- and S^{2-}, which promotes the formation of lots of Cu_2S precipitates. Thereby, the solubility of H_2S in the water film is further increased to promote the corrosion of copper by H_2S. Figure 14a–c are schematic diagrams of corrosion steps in LPG containing $H_2S + H_2O + O_2$. Combined with the previous theoretical analysis and Figure 14, it is shown that when both O_2 and H_2O exist in LPG, the corrosion grade of copper will be significantly increased. A large amount of Cu_2S and Cu_2O will be generated, which is consistent with the previous experimental results of the synergistic effect of humidity and O_2. Moreover, Cu_2O is thermodynamically more stable than Cu_2S (the standard free energies of formation at room temperature for Cu_2O and Cu_2S are −34.98 and −20.6 *kcal/mole*, respectively [27]), indicating that Cu_2O is preferentially formed at the same conditions. Therefore, the amounts of corrosion products in Cu_2O are more than Cu_2S. This conclusion is consistent with the results of XRD and EDS. In addition, a large amount of Cu_2O in XPS and FTIR also proves it.

$$O_{2(g)} \rightleftharpoons O_{2(l)} \quad (20)$$

$$O_{2\,(l)} + 2H_2O + 4e^- \rightleftharpoons 4\,OH^- \quad (21)$$

$$2Cu + 2OH^- \rightleftharpoons Cu_2O + H_2O + 2e^- \quad (22)$$

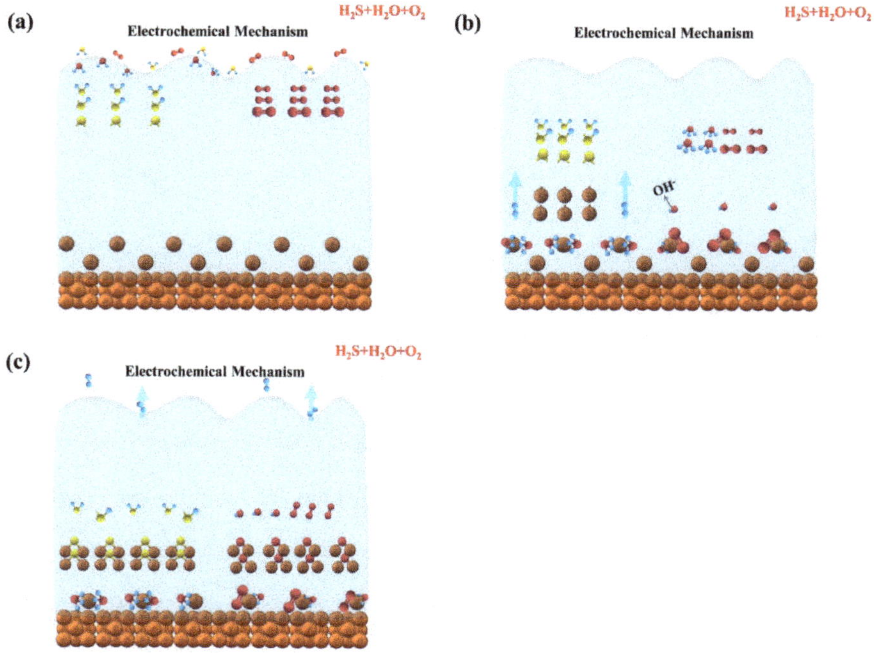

Figure 14. Schematic diagram of corrosion steps of copper sheet in LPG with H_2S + H_2O + O_2. (**a**) step1. (**b**) step2. (**c**) step3.

4. Conclusions

In this paper, the influence of humidity and O_2 on copper corrosion in H_2S-containing LPG was studied. The corrosion products were characterized and analysed to reveal the corrosion mechanism. The following conclusions were obtained:

1. In H_2S-containing LPG, *RH* has pronounced influence on the corrosion grade of copper. The variation in the *CP* with *RH* in LPG is a linear relationship.
2. The presence of O_2 in dry H_2S has limited influence on the corrosion of copper. The *CP* decreases linearly with the increase in O_2 concentration. In the presence of different *RHs*, the *CP* always follows a negative exponential function with O_2 concentration.
3. Surface morphologies of corrosion products obtained in different environments are quite different. Gas humidity and the presence of O_2 notably affect the microscopic morphology of corrosion products. In individual H_2S, the morphology of copper corrosion products is a regular hexagon block with sharp edges and corners. In H_2S + H_2O (100% *RH*), the morphology of copper corrosion products is uniform spherical shape. In H_2S + O_2, the morphology of copper corrosion products is irregular in shape and size. In H_2S + H_2O + O_2, the morphology of the corrosion products is a regular hexagon block with sharp edges and corners, spherical and irregular in shape and size.
4. In H_2S-containing LPG, *RH* and O_2 have obvious influence on the composition and distribution of corrosion products. In individual H_2S, the corrosion product of copper is only Cu_2S. In H_2S + H_2O, corrosion products of copper are mainly Cu_2S. In H_2S + O_2, corrosion products of copper are composed of a large amount of Cu_2S and a small amount of CuO and Cu_2O. In H_2S + H_2O + O_2, corrosion products are composed of a large amount of Cu_2O, Cu_2S and a small amount of CuO.
5. The corrosion mechanism of copper in LPG in the presence of different corrosive gases was proposed. The corrosive gas influences the corrosion mechanism remarkably. In individual H_2S and H_2S + O_2, the corrosion process is chemical in nature. H_2S and O_2

react with copper directly at the interface. The corrosion mechanism of copper in LPG containing $H_2S + H_2O$ and $H_2S + H_2O + O_2$ is an electrochemical corrosion process. In $H_2S + H_2O$, the corrosion proceeds with an anodic reaction of copper oxidization and a cathodic reaction of traditional hydrogen depolarization. In $H_2S + H_2O + O_2$, two different electrochemical reactions happen: one is the same as in $H_2S + H_2O$, and the other electrochemical reaction displays as the corrosion of O_2 in neutral medium, in which the cathodic process is oxygen depolarization.

Author Contributions: Conceptualization, H.W. and J.X.; methodology, H.W.; validation, H.W., X.L. and J.X.; formal analysis, Q.W.; investigation, Y.L.; writing—original draft preparation, X.L.; writing—review and editing, X.L.; supervision, H.W. All authors have read and agreed to the published version of the manuscript.

Funding: This research received no external funding.

Institutional Review Board Statement: Not applicable.

Informed Consent Statement: Not applicable.

Data Availability Statement: Not applicable.

Conflicts of Interest: The authors declare no conflict of interest.

References

1. Wang, B.W.; Jiang, Z.M.; Hu, X.C.; Guo, B.L. Discussion on corrosion control of LPG copper sheet in a terminal plant of Bohai oil field. *Chem. Manag.* **2015**, *21*, 3.
2. Yi, F.; Lu, Y.; Zhao, J.M. Influence and mechanism of sulfide in liquefied petroleum gas on copper corrosion. *J. Beijing Univ. Chem. Technol. (Nat. Sci. Ed.)* **2021**, *48*, 59–65.
3. Zhong, X.Y. *Research on Sulfur Corrosion and Removal of Light Oil Products*; China University of Petroleum: Beijing, China, 2011.
4. Halldin Stenlid, J.; Campos dos Santos, E.; Johansson, A.J.; Pettersson, L.G.M. Properties of interfaces between copper and copper sulphide/oxide films. *Corros. Sci.* **2021**, *183*, 109313. [CrossRef]
5. Long, Y.; Song, W.; Fu, A.; Xie, J.; Feng, Y.; Bai, Z.; Yin, C.; Ma, Q.; Ji, N.; Kuang, X. Combined effect of hydrogen embrittlement and corrosion on the cracking behaviour of C110 low alloy steel in O_2-contaminated H_2S environment. *Corros. Sci.* **2022**, *194*, 109926. [CrossRef]
6. Johnsen, R.; Lange, T.; Stenerud, G.; Olsen, J.S. Environmentally assisted degradation of spinodal copper alloy C72900. *Corros. Sci.* **2018**, *142*, 45–55. [CrossRef]
7. Dou, W.; Jia, R.; Jin, P.; Liu, J.; Chen, S.; Gu, T. Investigation of the mechanism and characteristics of copper corrosion by sulphate reducing bacteria. *Corros. Sci.* **2018**, *144*, 237–248. [CrossRef]
8. Echeverria, F.; Botero, C.A.; Correa, E.; Meza, D.; Castano, J.G.; Gomez, M.A. High resolution morphological changes of Cu, Ni, Al, and Au surfaces due to atmospheric corrosion. *IEEE Trans. Device Mater. Reliab.* **2017**, *17*, 331–339. [CrossRef]
9. Majtás, D.; Kreislová, K.; Turek, L. Failure of electric products by H_2S. *Koroze Ochr. Mater.* **2018**, *62*, 71–77. [CrossRef]
10. Zhu, Z.; Zuo, X.; Ying, Z. Corrosion analysis of copper T2 exposed to polluted atmospheres and study on prediction model. *Corros. Rev.* **2017**, *35*, 35–46. [CrossRef]
11. Odnevall, I. Atmospheric corrosion of copper in a rural atmosphere. *J. Electrochem. Soc.* **1995**, *142*, 3682–3689. [CrossRef]
12. Monzó, J.; García-Antón, J.; Guiñón, J.L. Influence of elemental sulphur and mercaptans on corrosion of copper strips in the ASTM D-130 test by means of electronic microscopy (SEM) and energy dispersive X-ray (EDX). *Fresenius J. Anal. Chem.* **1991**, *341*, 606–610. [CrossRef]
13. Monzó, J.; García-Antón, J.; Guiñón, J.L. Study of corrosion on copper strips by mixtures of mercaptans, sulphides and disulphides with elemental sulphur in the ASTM D-130 test by means of electron microscopy (SEM) and energy dispersive X-ray (EDX). *Fresenius J. Anal. Chem.* **1992**, *343*, 593–596. [CrossRef]
14. Toyoda, E.; Watanabe, M.; Higashi, Y.; Tanaka, T.; Miyata, Y.; Ichino, T. Depth profiling analysis of tarnish films on copper formed during short-term outdoor exposure by glow discharge optical emission spectroscopy. *Corrosion* **2004**, *60*, 729–735. [CrossRef]
15. Graedel, T.E.; Franey, J.P.; Gualtieri, G.J.; Kammlott, G.W.; Malm, D.L. On the mechanism of silver and copper sulfidation by atmospheric H_2S and OCS. *Corros. Sci.* **1985**, *25*, 1163–1180. [CrossRef]
16. Kong, D.; Dong, C.; Xu, A.; Man, C.; He, C.; Li, X. Effect of sulphide concentration on copper corrosion in anoxic chloride-containing solutions. *J. Mater. Eng. Perform.* **2017**, *26*, 1741–1750. [CrossRef]
17. Su, W.; Lv, W.; Liu, Z.; Zhang, Z. Corrosion of copper exposed to Zhanjiang and Zhuhai atmospheric environments. *Anti-Corros. Methods Mater.* **2017**, *64*, 286–292. [CrossRef]
18. Fujimoto, S.; Umemura, H.; Shibata, T. Atmospheric corrosion of electroplated cu thin film in moist oxygen environment. *ECS Trans.* **2006**, *1*, 243–247. [CrossRef]

19. Araban, V.; Kahram, M.; Rezakhani, D. Evaluation of copper atmospheric corrosion in different environments of Iran. *Corros. Eng. Sci. Technol.* **2016**, *51*, 498–506. [CrossRef]
20. FitzGerald, K.P.; Nairn, J.; Skennerton, G.; Atrens, A. Atmospheric corrosion of copper and the colour, structure and composition of natural patinas on copper. *Corros. Sci.* **2006**, *48*, 2480–2509. [CrossRef]
21. Kong, D.C.; Dong, C.F.; Fang, Y.H.; Xiao, K.; Guo, C.Y.; He, G.; Li, X.G. Copper corrosion in hot and dry atmosphere environment in Turpan, China. *Trans. Nonferrous Met. Soc. China* **2016**, *26*, 1721–1728. [CrossRef]
22. Lopesino, P.; Alcántara, J.; Fuente, D.; Chico, B.; Jiménez, J.; Morcillo, M. Corrosion of copper in unpolluted chloride-rich atmospheres. *Metals* **2018**, *8*, 866. [CrossRef]
23. Yan, F.; Li, X.; Jiang, B.; Lin, D.; Fu, M.; Li, W. Initial corrosion behaviour of pure copper in atmospheric environment. *IOP Conf. Ser. Earth Environ. Sci.* **2019**, *384*, 012039. [CrossRef]
24. Zhang, F.; Örnek, C.; Liu, M.; Müller, T.; Lienert, U.; Ratia-Hanby, V.; Carpén, L.; Isotahdon, E.; Pan, J. Corrosion-induced microstructure degradation of copper in sulphide-containing simulated anoxic groundwater studied by synchrotron high-energy X-ray diffraction and ab-initio density functional theory calculation. *Corros. Sci.* **2021**, *184*, 109390. [CrossRef]
25. Saha, D.; Pandya, A.; Singh, J.K.; Paswan, S.; Singh, D.D.N. Role of environmental particulate matters on corrosion of copper. *Atmos. Pollut. Res.* **2016**, *7*, 1037–1042. [CrossRef]
26. Hedin, A.; Johansson, A.J.; Lilja, C.; Boman, M.; Berastegui, P.; Berger, R.; Ottosson, M. Corrosion of copper in pure O_2-free water. *Corros. Sci.* **2018**, *137*, 1–12. [CrossRef]
27. Sharma, S.P. Reaction of copper and copper oxide with H_2S. *J. Electrochem. Soc.* **1980**, *127*, 21–26. [CrossRef]
28. Wu, T.; Zhou, Z.; Xu, S.; Xie, Y.; Huang, L.; Yin, F. A corrosion failure analysis of copper wires used in outdoor terminal boxes in substation. *Eng. Failure Anal.* **2019**, *98*, 83–94. [CrossRef]
29. Thethwayo, B.M.; Garbers-Craig, A.M. Laboratory scale investigation into the corrosion of copper in a sulphur-containing environment. *Corros. Sci.* **2011**, *53*, 3068–3074. [CrossRef]
30. Chen, J.; Qin, Z.; Martino, T.; Shoesmith, D.W. Non-uniform film growth and micro/macro-galvanic corrosion of copper in aqueous sulphide solutions containing chloride. *Corros. Sci.* **2017**, *114*, 72–78. [CrossRef]
31. Lu, X.; Liu, Y.; Liu, M.; Wang, Z. Corrosion behaviour of copper T2 and brass H62 in simulated Nansha marine atmosphere. *J. Mater. Sci. Technol.* **2019**, *35*, 1831–1839. [CrossRef]
32. Schindelholz, E.J.; Cong, H.; Jove-Colon, C.F.; Li, S.; Ohlhausen, J.A.; Moffat, H.K. Electrochemical aspects of copper atmospheric corrosion in the presence of sodium chloride. *Electrochim. Acta.* **2018**, *276*, 194–206. [CrossRef]
33. Deutscher, R.L.; Woods, R. Characterization of oxide layers on copper by linear potential sweep voltammetry. *J. Appl. Electrochem.* **1986**, *16*, 413–421. [CrossRef]
34. Tran, T.T.M.; Fiaud, C.; Sutter, E.M.M.; Villanova, A. The atmospheric corrosion of copper by hydrogen sulphide in underground conditions. *Corros. Sci.* **2003**, *45*, 2787–2802. [CrossRef]
35. Tran, T.T.M.; Fiaud, C.; Sutter, E.M.M. Oxide and sulphide layers on copper exposed to H_2S containing moist air. *Corros. Sci.* **2005**, *47*, 1724–1737. [CrossRef]
36. Nielsen, B.C.; Doğan, Ö.N.; Howard, B.H. Effect of temperature on the corrosion of Cu-Pd hydrogen separation membrane alloys in simulated syngas containing H_2S. *Corros. Sci.* **2015**, *96*, 74–86. [CrossRef]
37. Cai, L.; Chen, M.; Wang, Y.; Chen, C.; Zhang, L.; Zhou, H.; Wu, L.; Yan, Y. Electrochemical corrosion behaviour of bronze materials in an acid-containing simulated atmospheric environment. *Mater. Corros.* **2019**, *71*, 464–473. [CrossRef]
38. Yu, X.; Wang, Z.; Lu, Z. In situ investigation of atmospheric corrosion behaviour of copper under thin electrolyte layer and static magnetic field. *Microelectron. Reliab.* **2020**, *108*, 113630. [CrossRef]
39. Bellakhal, N.; Draou, K.; Brisset, J.L. Electrochemical investigation of copper oxide films formed by oxygen plasma treatment. *J. Appl. Electrochem.* **1997**, *27*, 414–421. [CrossRef]
40. *ASTM Standard D-130-80, Part 23*; American Society for Testing and Materials: West Conshohocken, PA, USA, 1981; p. 104.
41. Wang, H.; Xie, J.; Yan, K.P.; Duan, M.; Zuo, Y. The nucleation and growth of metastable pitting on pure iron. *Corros. Sci.* **2009**, *51*, 181–185. [CrossRef]
42. Persson, D.; Leygraf, C. In situ infrared reflection absorption spectroscopy for studies of atmospheric corrosion. *J. Electrochem. Soc.* **1993**, *140*, 1256–1260. [CrossRef]
43. Zhao, W.; Babu, R.P.; Chang, T.; Odnevall, I.; Hedström, P.; Johnson, C.M.; Leygraf, C. Initial atmospheric corrosion studies of copper from macroscale to nanoscale in a simulated indoor atmospheric environment. *Corros. Sci.* **2022**, *195*, 109995. [CrossRef]
44. Paul, A.M.; Sajeev, A.; Nivetha, R.; Gothandapani, K.; Bhardwaj, P.; Raghavan, V.; Jacob, G.; Sellapan, R.; Jeong, S.K.; Grace, A.N. Cuprous oxide/graphitic carbon nitride (g-C_3N_4) nanocomposites for electrocatalytic hydrogen evolution reaction. *Diamond Relat. Mater.* **2020**, *107*, 107899. [CrossRef]
45. Shanghai Institute of Organic Chemistry of CAS. Chemistry Database. 1978–2022. Available online: http://www.organchem.csdb.cn. (accessed on 1 February 2022).
46. Li, K.; Chen, Z.; Li, J.; Sun, X.; Xu, F.; Xu, L. Corrosion mechanism of copper immersed in ammonium sulphate solution. *Mater. Corros.* **2018**, *69*, 1597–1608. [CrossRef]
47. Chialvo, M.R.G.; Marchiano, S.L.; Arvía, A.J. The mechanism of oxidation of copper in alkaline solutions. *J. Appl. Electrochem.* **1984**, *14*, 165–175. [CrossRef]

metals

Article

Numerical Investigation on the Effect of Residual Stresses on the Effective Mechanical Properties of 3D-Printed TPMS Lattices

Nissar Ahmed [1], Imad Barsoum [1,2,*] and Rashid K. Abu Al-Rub [1,2]

1 Advanced Digital & Additive Manufacturing Center, Khalifa University, Abu Dhabi P.O. Box 127788, United Arab Emirates
2 Department of Mechanical Engineering, Khalifa University, Abu Dhabi P.O. Box 127788, United Arab Emirates
* Correspondence: imad.barsoum@ku.ac.ae

Abstract: The layer-by-layer process of additive manufacturing (AM) is known to give rise to high thermal gradients in the built body resulting in the accumulation of high residual stresses. In the current study, a numerical investigation is conducted on the effect of residual stresses on the mechanical properties of IN718 triply periodic minimal surface (TPMS) lattices fabricated using the selective laser melting (SLM) process for different relative densities. The AM simulation of four different sheet- and ligament-based TPMS topologies, namely, Schwarz Primitive, Schoen Gyroid, Schoen IWP-S, and IWP-L, are performed using a sequentially coupled thermomechanical finite element model to evaluate the thermal histories and residual stress evolution throughout the SLM process. The finite element results are utilized to obtain the effective mechanical properties, such as elastic modulus, yield strength, and specific energy absorption (SEA), of the TPMS lattices while accounting for the residual stress field arising from the SLM process. The mechanical properties are correlated to relative density using the Gibson–Ashby power laws and reveal that the effect of the residual stresses on the elastic modulus of the as-built TPMS samples can be significant, especially for the Schwarz Primitive and Schoen-IWP-L TPMS topologies, when compared to the results without accounting for residual stresses. However, the effect of the residual stresses is less significant on yield strength and SEA of the TPMS samples. The work demonstrates a methodology for numerical simulations of the SLM process to quantify the influence of inherited residual stresses on the effective mechanical properties of complex TPMS topologies.

Keywords: residual stresses; additive manufacturing; selective laser melting (SLM); finite element modeling (FEM); triply periodic minimal surface (TPMS)

Citation: Ahmed, N.; Barsoum, I.; Abu Al-Rub, R.K. Numerical Investigation on the Effect of Residual Stresses on the Effective Mechanical Properties of 3D-Printed TPMS Lattices. *Metals* **2022**, *12*, 1344. https://doi.org/10.3390/met12081344

Academic Editor: Zhanyong Zhao

Received: 29 June 2022
Accepted: 8 August 2022
Published: 12 August 2022

1. Introduction

Over the past couple of decades, advancements in additive manufacturing (AM) have enabled the fabrication of complex lattice structures including the triply periodic minimal surface (TPMS) [1,2], but these structures are difficult to produce with conventional manufacturing methods due to the associated high cost and low efficiency [3]. TPMS structures are three-dimensional open-celled structures composed of one or more repeating unit cells in an orderly pattern [4]. These structures have a high specific strength and stiffness, and a high surface-area-to-volume ratio, which are features desired in many mechanical applications [5,6]. However, due to the sequential layer-by-layer nature of the selective laser melting (SLM) process, part of the heat is absorbed during melting of the powder layer and part of the heat is conducted to the already solidified layers below and through convection to the surroundings. This cyclic nature of thermal loads leads to a transient change in temperature. The steep thermal gradients induce a strain mismatch between the newly formed layers and pre-solidified layers beneath during heating and cooling cycles, resulting in the accumulation of residual stresses [7]. The inherited residual stresses in SLM are known to affect the dimensional accuracies and shape of the parts, contributing to crack formation and delamination of layers. Additionally, residual stress drastically affects the fatigue life and the loading capacity of the parts [8,9]. Hence, quantifying residual stresses arising from the SLM process is becoming more critical. Depending on the magnitude of the residual stresses and their effect, they can be classified as type I, II, or III residual stresses [10]. Type I residual stresses act over larger length scales, e.g., macroscale, due to nonuniform heating and cooling rates throughout the AM process, which can result in non-negligible deformations. Type II stresses that act at the microstructural level, e.g., grain size level, can occur due to phase transformation in the material at the microscale, while type III stresses exist over atomic scales caused by dislocation and point defects. The focus of the current study is macroscopic residual stresses of type I. As the SLM process involves many process parameters, experimentally evaluating residual stresses is time-consuming [11,12], and hence, numerical analysis based on the finite element method (FEM) is an efficient way simulate the thermomechanical behavior and quantify the residual stresses during the SLM process. The most widely used thermomechanical approach is the indirect sequentially coupled analysis that is performed in two stages. First, a transient temperature field of the built part is simulated, and the temperature results are then used to perform the mechanical analysis. This method is computationally less expensive compared to fully coupled analysis in which the temperature and displacement degrees of freedom are solved simultaneously and updated for each time increment [13]. An indirect coupling approach has been successfully used to investigate the AM process to predict residual stress formations and geometric deformations to good accuracies [14–19].

Numerical methods based on finite element analysis (FEA) have been extensively used alongside experiments to investigate the mechanical properties of TPMS structures fabricated from the AM process [20–24]. Al-Rub et al. [25] used FEA to investigate the effective anisotropic elastic and plastic properties along with the deformation mechanism under different load combinations of Schoen's IWP TPMS and proposed a cost-effective finite-element simulation framework for the IWP structural system. Al-Ketan et al. [26] studied the topology–property relationship of lattice structures based on the TPMS IWP minimal surfaces, and FEA was used to investigate the stiffness and strength of different relative densities of the TPMS structure along with experimental investigations. The work showed that the sheet-based IWP lattice structure exhibited high structural efficiency compared to other strut-based and skeletal-based lattice structures. Yang et al. [27] used a numerical approach to quantify the influence of various geometric factors such as surface thickness, sample size, and number of surface periods, etc., on the overall structural response of gyroid structures. Lee et al. [28] investigated the mechanical properties and deformation behavior under several different stress states of the Schwarz Primitive unit cell under periodic boundary conditions. Zheng et al. [29] using the finite volume method found that the mechanical properties are highly dependent on topological architectures of TPMS structures. Abueidda et al. [30] in their work predicted and compared the electrical/thermal conductivities, elastic properties, and anisotropies of different TPMS foams. Maskery et al. [31] used FEA to predict the stress distribution in polymer-based TPMSs under compressive loads with good agreement with experiments, and the investigation also indicated that the cell geometry played a key role in determining the deformation mechanism, failure modes, and the associated mechanical properties.

In general, FEA has been effectively used to investigate the elastic and plastic properties of TPMS structures; however, to the best of the authors knowledge, numerical thermomechanical analysis to quantify the effect of residual stresses inherited from the SLM process on the effective mechanical properties of TPMS structures is a topic not addressed in the literature. Hence, in this work, an FEA simulation scheme is proposed to evaluate the residual stresses from the SLM process and to quantify its influence on the effective mechanical properties of TPMS structures. Four different TPMS topologies are considered: Schwarz sheet-based Primitive, Schoen sheet-based Gyroid, Schoen sheet-based I-WP, and ligament-based I-WP (henceforth referred to P-S, G-S, IWP-S, and IWP-L, respectively). Inconel 718 (IN718), which is a commonly used grade for metallic 3D printing, is considered as the base material, and the effective mechanical properties from the thermomechanical simulations of the AM-built TPMS lattices are compared with their reference counterparts, i.e., lattices without residual stresses.

2. Modeling Framework

Figure 1 shows a schematic representation of the AM process with governing thermal conditions. Assuming a homogenous medium with isotropic thermal properties, e.g., no spatial change in thermal conductivity of the material, the transient temperature distribution $T\ (x, y, z, t)$ throughout the workpiece during the SLM process is governed by the following three-dimensional thermal transient conduction equation:

$$k(T)\ \nabla^2 T + Q = \rho\ C_p(T)\ \frac{\partial T}{\partial t} \tag{1}$$

where C_p is the temperature-dependent specific heat capacity, ρ is the density, T is the temperature, t is the time, Q is the internal heat generation rate, and k is the temperature-dependent thermal conductivity of the isotropic material. The initial and final (e.g., cooling stage) thermal condition for temperature distribution throughout the powder bed is given by

$$T(x, y, z, t = 0) = T_0,\ T(x, y, z, t = \infty) = T_0 \tag{2}$$

where T_0 = 22 °C corresponds to room temperature. The base of the substrate is preheated and subjected to a temperature of 100 °C during the build stage and is brought back to room temperature during the cooling phase. The boundary conditions for all other surfaces are assumed to be under heat loss to the surrounding gas by free air convection with a heat transfer coefficient, $h = 1 \times 10^{-5}$ W/mm² °C.

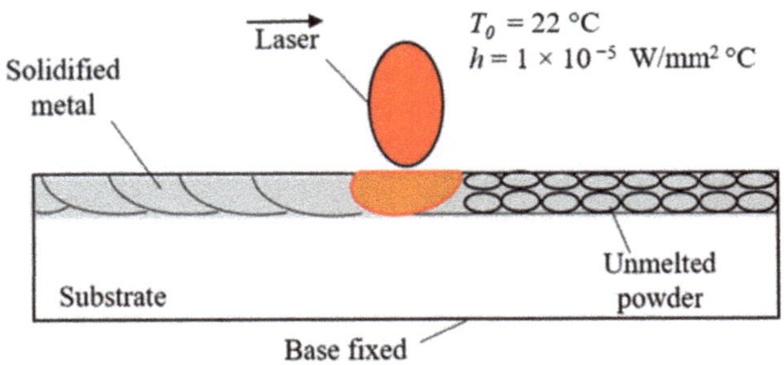

Figure 1. Representation of the AM process with thermal conditions.

The layer-by-layer formation during the SLM process gives rise to large thermal gradients in parts, which, in turn, leads to the accumulation of residual stress and strain during the solidification phase. The related stress and strain in the parts are associated with the following equation:

$$\{\sigma\} = [D]\{\varepsilon^e\} \tag{3}$$

where $\{\sigma\}$ is the stress vector, $[D]$ is the elasticity stiffness matrix, and $\{\varepsilon^e\}$ is the elastic strain vector. Using a simplified elastic–plastic hardening model, $\{\varepsilon^e\}$ can be expressed as:

$$\{\varepsilon^e\} = \{\varepsilon\} - \{\varepsilon^p\} - \{\varepsilon^t\} \tag{4}$$

where ε is the total strain vector, $\{\varepsilon^p\}$ is the plastic strain vector, and $\{\varepsilon^t\}$ is the thermal strain vector. Using Equation (4) in Equation (3) may be written as follows:

$$\{\varepsilon\} = [D]^{-1}\{\sigma\} + \{\varepsilon^p\} + \{\varepsilon^t\} \tag{5}$$

For isotropic material, the above stress–strain relationship can be elaborated in Cartesian co-ordinates as follows [32]:

$$\varepsilon_x = \frac{1}{E}\left[\sigma_x - v\left(\sigma_y + \sigma_z\right)\right] + \varepsilon_x^p + \varepsilon^t \tag{6}$$

$$\varepsilon_y = \frac{1}{E}\left[\sigma_y - v\left(\sigma_x + \sigma_z\right)\right] + \varepsilon_y^p + \varepsilon^t \tag{7}$$

$$\varepsilon_z = \frac{1}{E}\left[\sigma_z - v\left(\sigma_x + \sigma_y\right)\right] + \varepsilon_z^p + \varepsilon^t \tag{8}$$

$$\gamma_{xy} = \frac{\tau_{xy}}{2G} + \gamma_{xy}^p, \quad \gamma_{xz} = \frac{\tau_{xz}}{2G} + \gamma_{xz}^p, \quad \gamma_{yz} = \frac{\tau_{yz}}{2G} + \gamma_{yz}^p \tag{9}$$

where E is the elastic modulus; G is the shear modulus; ν is Poisson's ratio; τ_{xy}, τ_{xz}, and τ_{yz} are the shear stress components; γ_{xy}, γ_{xz}, and γ_{yz} are the corresponding shear strain components. The thermal strain component arising due to volume change caused by temperature variations can be expressed as follows:

$$\varepsilon^t = \alpha_e \Delta T = \alpha_e\left(T - T_{ref}\right) \tag{10}$$

where α_e is the coefficient of thermal expansion, T is the instantaneous temperature, and T_{ref} is the reference temperature with respect to time at $t = 0$. The deviator stresses according to the Prandtl–Reuss equation of plasticity can be represented as follows:

$$\frac{d\varepsilon_x^p}{\sigma'_x} = \frac{d\varepsilon_y^p}{\sigma'_y} = \frac{d\varepsilon_z^p}{\sigma'_z} = \frac{d\gamma_{xy}^p}{\tau_{xy}} = \frac{d\gamma_{yz}^p}{\tau_{yz}} = \frac{d\gamma_{zx}^p}{\tau_{zx}} = d\lambda \tag{11}$$

$$\sigma'_x = \sigma_x - \sigma_m, \qquad \sigma'_y = \sigma_y - \sigma_m, \qquad \sigma'_z = \sigma_z - \sigma_m \tag{12}$$

where σ'_x, σ'_y, and σ'_z are the deviator stresses in the x, y, and z directions, respectively; $d\lambda$ is the instant positive constant of proportionality. Σ_m refers to the mean stress, which can be evaluated as follows:

$$\sigma_m = \frac{\sigma_x + \sigma_y + \sigma_z}{3} \tag{13}$$

Substituting the values of ε_x^p, ε_y^p, ε_z^p, and ε^t in Equations (6)–(9), the resultant equations can be stated as below in Equations (14)–(17):

$$\varepsilon_x = \frac{1}{E}\left[\sigma_x - v\left(\sigma_y + \sigma_z\right)\right] + \int \sigma'_x d\lambda + \alpha_e \Delta T \tag{14}$$

$$\varepsilon_y = \frac{1}{E}\left[\sigma_y - v\left(\sigma_x + \sigma_z\right)\right] + \int \sigma'_y d\lambda + \alpha_e \Delta T \tag{15}$$

$$\varepsilon_z = \frac{1}{E}\left[\sigma_{zy} - v\left(\sigma_x + \sigma_y\right)\right] + \int \sigma'_z d\lambda + \alpha_e \Delta T \tag{16}$$

$$\gamma_{xy} = \frac{\tau_{xy}}{2G} + \int \tau_{xy} d\lambda, \; \gamma_{xz} = \frac{\tau_{xz}}{2G} + \int \tau_{xz} d\lambda, \; \gamma_{yz} = \frac{\tau_{yz}}{2G} + \int \tau_{yz} d\lambda \tag{17}$$

Finally, the Von Mises stress σ_m is computed as follows:

$$\sigma_m = \sqrt{\frac{1}{2}\left[\left(\sigma_x - \sigma_y\right)^2 + \left(\sigma_y - \sigma_z\right)^2 + \left(\sigma_z - \sigma_x\right)^2 + 6\left(\tau_{xy}^2 + \tau_{yz}^2 + \tau_{zx}^2\right)\right]} \tag{18}$$

where σ_x, σ_y, and σ_z are the x, y, and z components of stress, respectively.

2.1. Material

Temperature-dependent physical and thermal properties of Inconel 718 (IN718) including density ρ, specific heat C_p, and thermal conductivity k are used as input properties to perform the transient thermal analysis. For the transient mechanical analysis, the temperature-dependent elastic modulus E, yield strength σ_y, Poisson's ratio ν, plastic tangent modulus E_T, and coefficient of thermal expansion α are used.

The stress–strain curves of IN718 at various temperatures are shown Figure 2a, pertaining to stress normalized with the yield strength (σ_{y0}) at room temperature (T_0). The evolution of the temperature-dependent thermal, physical, and mechanical properties is shown in Figure 2b,c, respectively, where the properties are normalized with their corresponding value at room temperature, and the temperature is normalized with the melting temperature value of IN718 (e.g., T_m = 1260 °C). The corresponding properties at room temperature for IN718 are provided in Table 1.

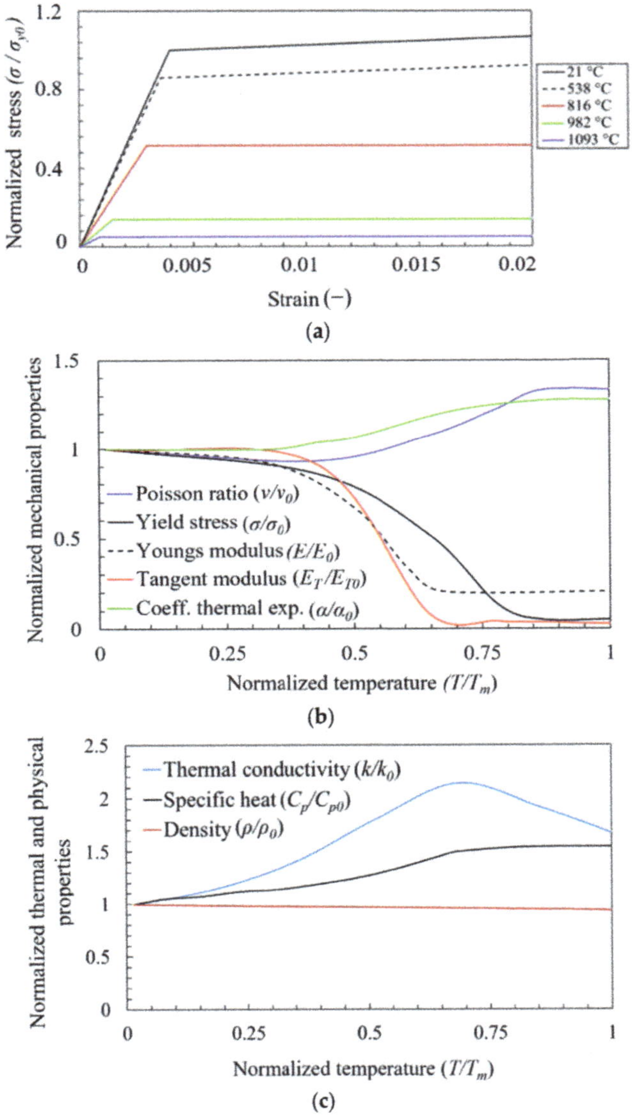

Figure 2. Normalized temperature-dependent properties for IN718: (**a**) stress–strain curves, (**b**) mechanical properties, and (**c**) thermal and physical properties.

Table 1. IN718 physical, mechanical, and thermal material properties at room temperature (T_0) [33].

Property	Value
Density ρ_0 (kg/m^3)	8220
Specific heat C_{p0} (J/kg °C)	421
Conductivity k_0 (W/m °C)	11.9
Yield strength σ_0 (MPa)	648
Poisson ratio ν_0	0.3
Thermal expansion α_0 (°/C)	1.44×10^{-5}
Young's modulus E_0 (GPa)	165

2.2. Design of Triply Periodic Minimum Surfaces (TPMS)

In TPMS structures, the mean curvature of a minimal surface is zero at every point and periodic in the three perpendicular directions. Table 2 provides the level-set approximations for the TPMS surfaces used in this work (P-S for Schwartz Primitive, G-S for Schoen Gyroid, and IWP for Schoen I-wrapped package) for generating the lattices in terms of local Cartesian coordinates, x, y, and z, and a specified level-set constant, c. A detailed design process for creating TPMS topologies was covered in the work by Al-Ketan and Abu Al-Rub [4]. Furthermore, Al-Ketan and Abu Al-Rub [34] developed a free software called MSLattice for generating TPMS-based lattices of either sheet-network type or ligament-network type. In sheet-network TPMS lattices, the level-set parameter c is set to zero such that the minimal surface splits the 3D space into two domains of equal volumes, and the relative density is varied through the thickening of the minimal surface, whereas, in ligament-network TPMS lattices, the level-set parameter c is used to control the proportion volumes of the two domains split by the minimal surface such that the smaller volume is solidified (e.g., for $c = 0$, the relative density is 50%). For this study, TPMS lattice structures shown in Figure 3 are investigated, where P-S, G-S, and IWP-S are sheet-based TPMS structures, while IWP-L is a ligament-based TPMS lattice. Three relative densities ($\overline{\rho} = \rho/\rho_s$) of values 10%, 20%, and 30% of the mentioned TPMS structure are considered for this study, where ρ is the apparent density of the lattice structure and ρ_s is the density of its base solid material. The $3 \times 3 \times 3$ cell configuration is selected for this study based on the yield stress convergence studies performed using different cell configurations (e.g., $3 \times 3 \times 3$, $4 \times 4 \times 4$, $5 \times 5 \times 5$, and $6 \times 6 \times 6$) considered for the P-S ($\overline{\rho} = 20\%$) lattice structure, with the method of evaluating yield strength as described in Section 2.4. As shown in Figure 4, the normalized yield stress value of different cell configurations is compared with that calculated from the equation by Lee et al. [28], which uses a unit cell P-S lattice structure with periodic boundary conditions and assumes the base material to be an isotropic solid with elastic modulus E_s = 200 GPa, yield strength σ_s = 400 MPa, and Poisson's ratio ν_s = 0.3. Although it is observed that a higher cell configuration of $6 \times 6 \times 6$ gives a value closer to matching the results by Lee et al. [28], the time required to simulate such an AM build configuration will be very large and with only a marginal gain in accuracy. Hence, to reduce the overall computational time, the $3 \times 3 \times 3$ cell configuration is used for simulating residual stresses of the TPMS structures as well as their effective mechanical properties, which is within a 2% agreement with the results in [28]. The TPMS lattices are designed using the MSLattice, Al-Ketan and Abu Al-Rub (Abu Dhabi, UAE) simulation tool [34].

Table 2. Level-set equation for P-S, G-S, and IWP [4], where x, y, and z are Cartesian coordinates, and c is the level-set constant that controls the thickness of TPMS surfaces.

TPMS	Level-Set Equations
P-S	$\cos x + \cos y + \cos z = c$
G-S	$\sin x \cos y + \sin y \cos z + \sin z \cos x = c$
IWP	$2\,(\cos x \cos y + \cos y \cos z + \cos z \cos x) - (\cos 2x + \cos 2y + \cos 2z) = c$

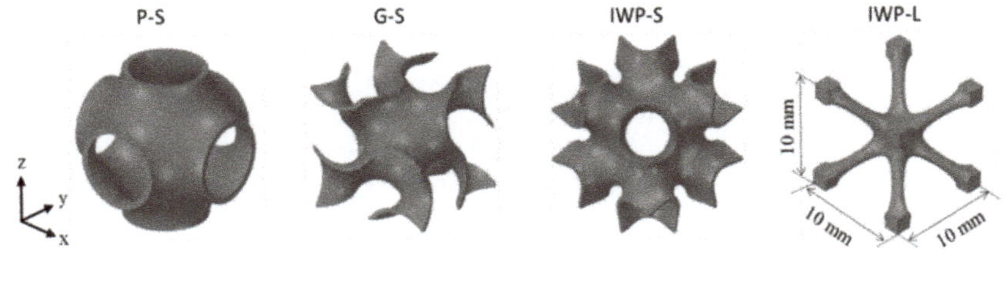

Figure 3. TPMS unit cell topologies pertaining to, e.g., relative density of $\bar{\rho}$ = 10, with unit cell dimensions of 10 mm × 10 mm × 10 mm.

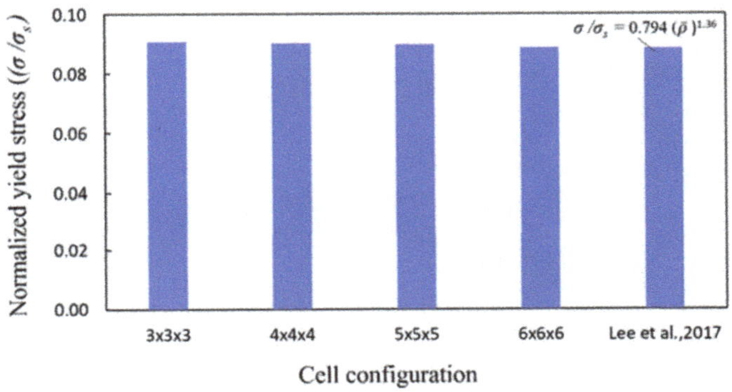

Figure 4. Convergence study of normalized yield strength values for different cell configurations of P-S ($\bar{\rho}$ = 20%) compared with calculated value from Lee et al. [28].

2.3. Residual Stress Simulation

A nonlinear transient thermal analysis of the layer-by-layer buildup of the process is first performed to obtain the temperature histories during the melting and solidification stages. This is followed by a transient elastic–plastic mechanical analysis, where the transient temperature profile from the transient thermal analysis is used as input in the mechanical model. Ansys-WB [33] provides the multiphysics capability for performing such sequentially coupled thermomechanical analysis. The internal residual stress formation occurs during the cyclic heating and cooling due to the layer-by-layer nature of the SLM process and continually evolves during the process. The residual stress profile throughout the part is attained after the final cooling phase to room temperature (e.g., 22 °C). The baseplate removal option available in the simulation tool [33] is performed and the final residual stress state of the build is used as the initial stress profile for evaluating the effective mechanical properties of the TPMS samples.

The processing parameters used for performing the TPMS AM build simulation with IN718 material are as follows: the laser scan speed is 1200 mm/s, the powder layer thickness is 40 μm, and the preheat temperature of the IN718 base plate is set to 100 °C and switches to room temperature during the cooling phase after completion of the build, where build supports are not considered for this study. The simulation is performed on the four TPMS topologies considering three different relative densities, e.g., $\bar{\rho}$ = 10%, 20%, and 30%. The AM simulation tool performs a thermal analysis followed by a quasi-static elastic–plastic mechanical analysis. The thermal analysis records the temperature histories of elements activated layer-wise using the element birth and death technique until completion of the build and cooling to room temperature; this layer-wise activation is illustrated in Figure 5, e.g., the first element layer (highlighted in purple) is activated for deposition while the rest of the element layers above represent the material to be deposited and are to be activated in subsequent stages. The thermal histories are then used as input for mechanical analysis to evaluate the stresses by layer-wise activation in a similar pattern as in the thermal analysis, and the stress state at the cooling of the build is calculated. As modeling every physical layer of deposition is computationally expensive and impractical, a simplified lumped-layer approach is used in which activation of each finite element layer accounts for the deposition of several layers of actual metal powder at once [33]. Incorporating this approach, the TPMS lattice is modeled with hexahedral elements (e.g., HEX8) with an element layer height of 0.4 mm as per the size range (i.e., 10~20 times the powder layer) [33]; thus, for this work, a total build height of 30 mm will have 75 element activation layers corresponding to 750 physical layers of actual metal powder, where the base plate (50 mm × 50 mm × 10 mm) uses a coarser mesh size of 5 mm. The meshed models for the four TPMS lattice structures are shown in Figure 6, pertaining to relative density $\overline{\rho}$ = 10%, with a total number of elements in the range of 250,000 to 600,000 for different relative densities of the TPMS lattice.

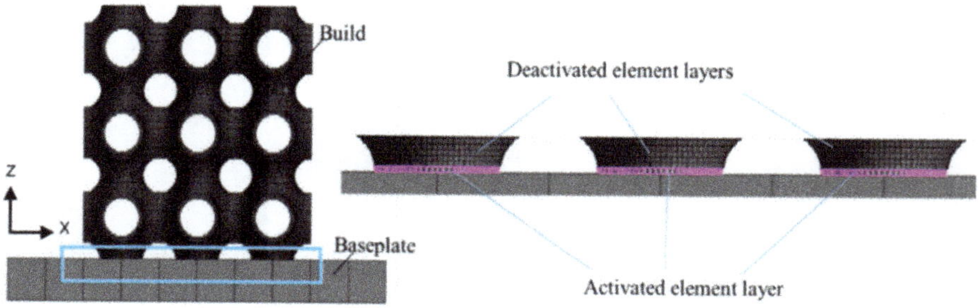

Figure 5. The illustration shows the activation of the first layer of elements (highlighted in purple color) for deposition; the rest of the layers above are in the deactivated state representing the material to be deposited.

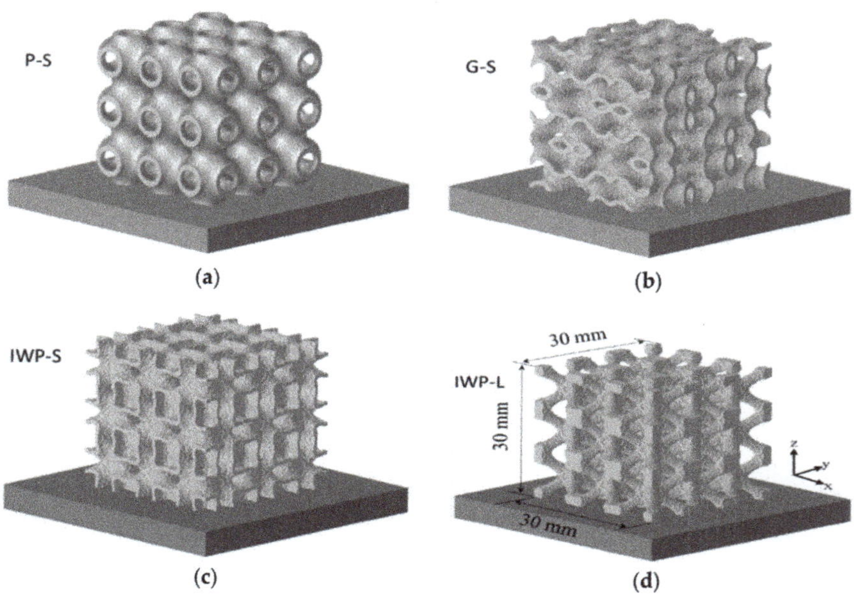

Figure 6. 3 × 3 × 3 cell configuration FEA models used for AM buildup simulation on an IN718 baseplate: (**a**) P-S (**b**) G-S (**c**) IWP-S, and (**d**) IWP-L.

The various stages of build formation during the thermal analysis are consolidated and shown in Figure 7a–f pertaining to the P-S lattice structure ($\overline{\rho}$ = 10%) through the built stage, e.g., build height z_{bh}. The thermal analysis terminates after the cooling phase when the part cools to room temperature (22 °C). Similarly, layer-by-layer simulation is performed during transient structural analysis using temperature histories, and the final vector principal stress on the nodes is grouped and exported to be used as an input as pre-stress to evaluate the mechanical properties during tensile and compression loading simulation.

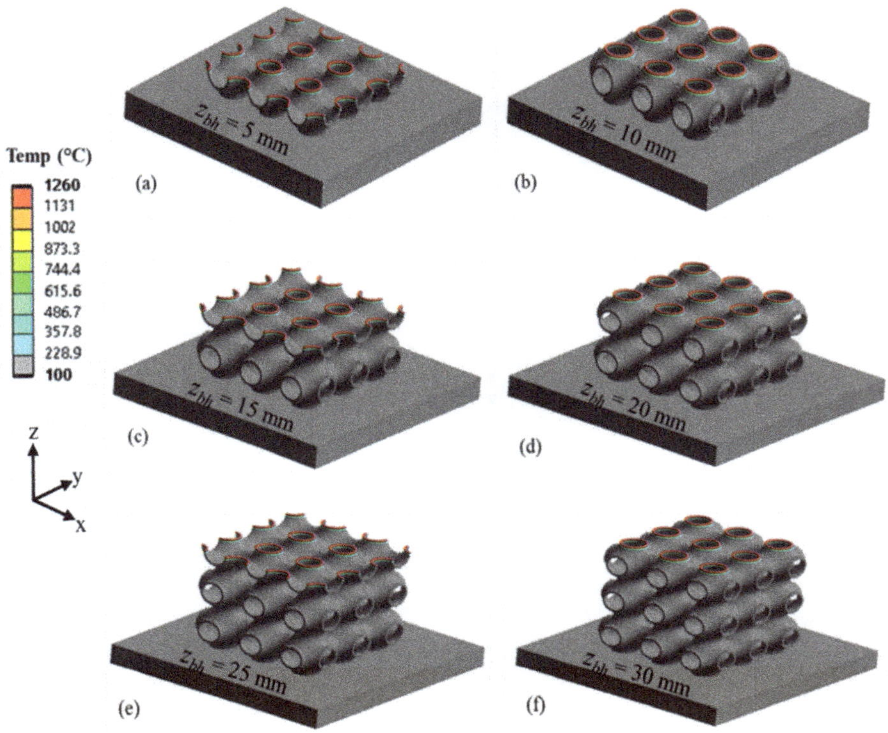

Figure 7. Various stages of 3 × 3 × 3 AM buildup simulation for P-S ($\overline{\rho}$ =10%) for build height: (**a**) z_{bh} = 5 mm, (**b**) z_{bh} = 10 mm, (**c**) z_{bh} = 15 mm, (**d**) z_{bh} = 20 mm, (**e**) z_{bh} = 25 mm, and (**f**) z_{bh} = 30 mm at completion of build.

2.4. Mechanical Simulation

The final stress state evaluated from the AM simulation is imported as the initial residual stresses field for investigating the mechanical properties of the corresponding TPMS lattices under tensile and compressive loading. The mechanical properties of the TPMS are determined considering uniaxial tensile and compression loading in the build direction (z-direction). The finite element model uses a tetrahedral element with a mesh size of 0.25 mm from mesh convergence studies performed for estimating the yield strength of the configuration of the P-S ($\overline{\rho}$ = 20%) lattice structure, and a bilinear elastic-plastic material model with linear plastic hardening is used for all mechanical simulations. Figure 8 shows the representation of the tensile and compression loading case and symmetric boundary conditions in the z-direction applied on the base plane for a P-S lattice structure. In the mechanical simulation, all six components of the stress tensor are mapped onto the nodes of the FE model as initial conditions. A displacement load in the z-direction is applied to the top surface of the lattice, with a magnitude of $\Delta = \pm 3$ mm in tension or compression, respectively, which corresponds to a total normal average strain of 10%. This is applied via a remote reference point coupled to the top built surface [33]. The reaction force at the remote node is monitored to obtain the uniaxial stress–strain response during the loading of the TPMS lattices, where the uniaxial stress is defined as the ratio between the reaction force and the area of the TPMS lattice (30 mm × 30 mm). The yield strength is determined from the stress–strain response through the 0.2% offset strain method, whereas the elastic modulus in tension and compression is evaluated from the slope of the stress–strain curves in the elastic region. The uniaxial tensile and compression loading simulations are also performed incorporating the effect of the full-field residual stress profile imported from the AM simulation (c.f., Figure 10) and the results are compared with the case without residual stresses as the initial condition for all the TPMS lattices and relative densities.

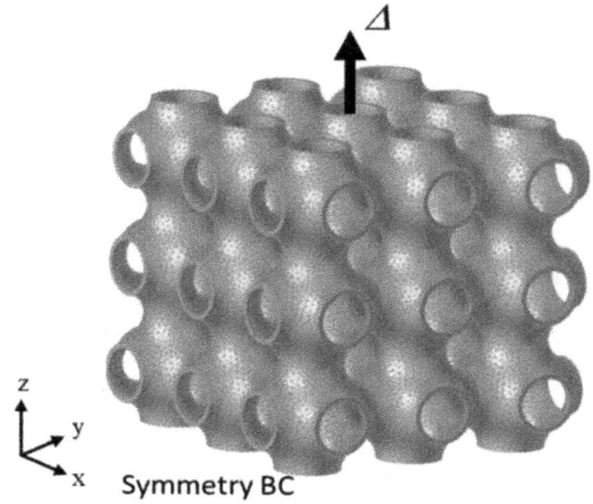

Figure 8. FE model of P-S pertaining to $\overline{\rho}$ = 10% with uniaxial loading and symmetric BC.

3. Results and Discussion

To verify the modeling framework proposed herein and the residual stress results, the current simulation approach of layer-wise activation described in Section 2.3 is compared with a case of experimentally measured residual stress by An et al. [35] on a curved thin-walled plate, for which the details are presented in Appendix A.

For the TPMS lattice structures and for their three relative densities, the evolution of stresses for the total build height of the samples is shown in Figure 9. The stress is calculated by determining the reaction force of the constrained base nodes in the build direction (z-direction) over its projected base area (30 mm × 30 mm). This provides a measure of the level of the stress in the build direction, which, as can be seen in Figure 9, is compressive in nature during the build-up stage, with an increase in magnitude as the relative density increases. The stress level during the build-up stage is observed to be higher in G-S and IWP-S lattice structures as compared to P-S and IWP-L for the same densities. In Figure 10, which shows the Von Mises residual stress contours for all TPMS lattices at relative density $\overline{\rho}$ =10%, the color contours are unified and used within the same limits for comparison purposes. As evident, the residual stress profiles differ for each TPMS, and hence, its effect on the effective mechanical properties will vary for each topology. It is also interesting to note that the residual stresses are nonuniformly distributed through the printing z-direction. The predicted stress values are higher in magnitudes than the initial yield stress of the material (e.g., 648 MPa). This is expected as the finite element model used in the current study does not consider the convective heat transfers in the molten pool; hence, computed peak temperatures and temperature gradients are slightly overestimated and consequently higher residual stress values are determined [36].

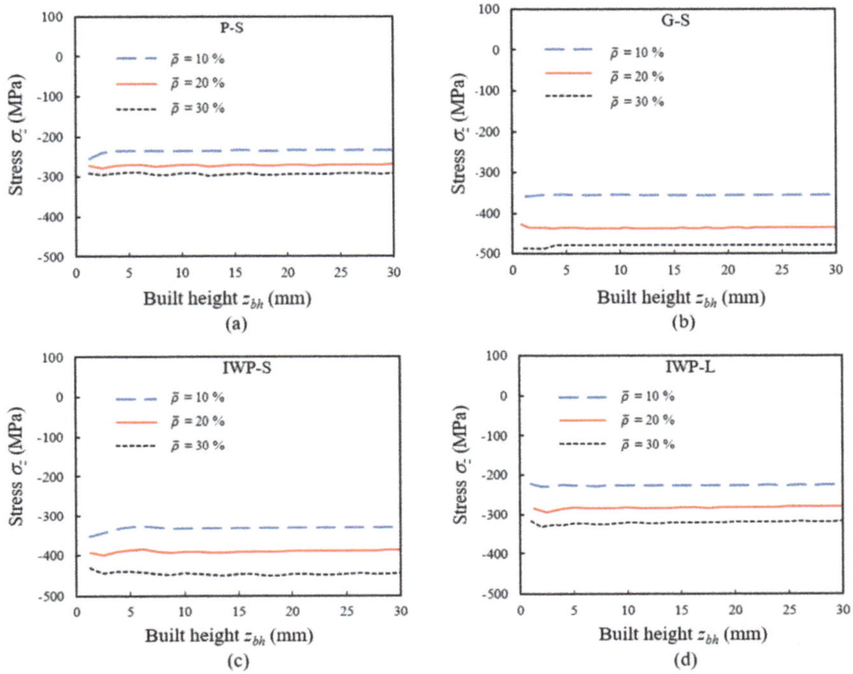

Figure 9. Stress evolution in build direction during the build height (30 mm) of TPMS for various relative densities: (**a**) PS, (**b**) G-S, (**c**) IWP-S, and (**d**) IWP-L.

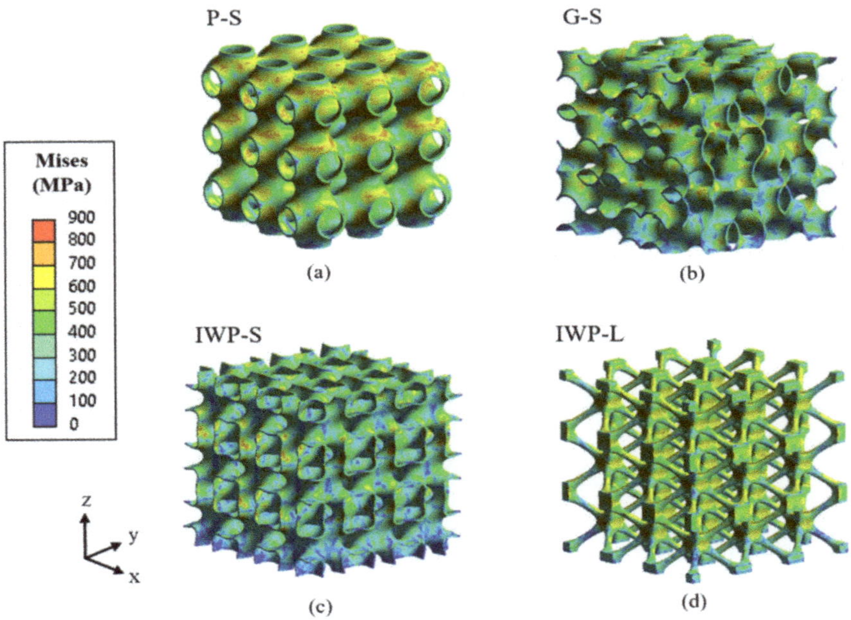

Figure 10. Von Mises residual stress field in the TPMS lattices build of $\bar{\rho}$ = 10%: (**a**) P-S, (**b**) G-S, (**c**) IWP-S, and (**d**) IWP-L.

3.1. Stress–Strain Curves

The uniaxial tensile and compression stress–strain curves for all four TPMS lattices and the three relative densities considered are shown in Figures 11–13, both for the cases with and without the incorporation of residual stress effects from the AM process. As can be seen for P-S in Figures 11a, 12a and 13a, the effect of the residual stress on both stiffness and onset of yielding is notable, especially under compression loading. This is attributed to the high level of residual stresses present in the P-S topology, as shown in Figure 10a, as compared to the G-S and IWP-S in Figure 10b,c, respectively. A similar trend in the influence of the residual stresses on the mechanical response can be observed in Figures 12d and 13d for IWP-L, which is also attributed to the level of residual stresses associated with the IWP-L topology, as shown in Figure 10d. For all the sheet-based TPMS topologies, e.g., P-S, G-S, and IWP-S, subjected to compression loading, e.g., Figure 11a–c, softening is observed beyond the point of onset of yielding, especially for $\overline{\rho}$ =10%, which is associated with localized deformation and buckling of the TPMS walls. This trend diminishes as $\overline{\rho}$ increases; however, the difference in mechanical response between the compression and tension loading persists for all $\overline{\rho}$. At large deformations, e.g., strain values larger than 2%, it is observed that the mechanical response of all the TPMS topologies and for all relative densities is not affected by the presence of residual stress, neither for compression nor tension loading.

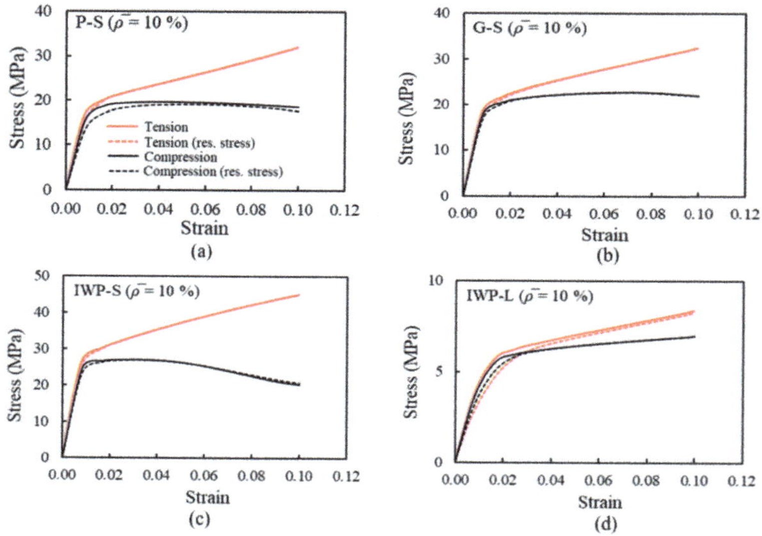

Figure 11. The stress–strain curve for tension/compression loading for $\overline{\rho}$ = 10% with and without residual stress: (**a**) P-S, (**b**) G-S, (**c**) IWP-S, and (**d**) IWP-L.

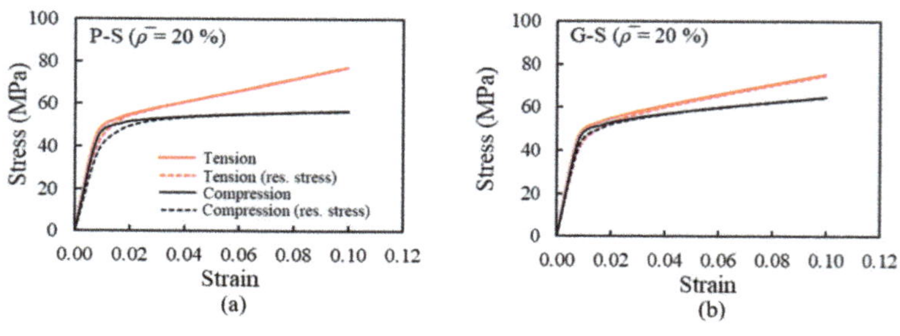

Figure 12. *Cont.*

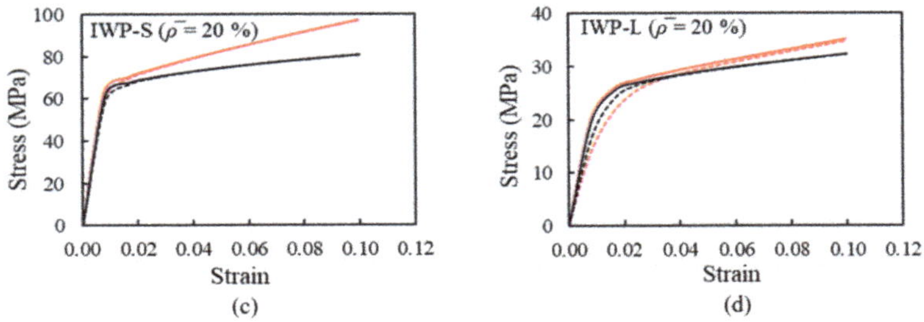

Figure 12. The stress–strain curve for tension/compression loading for $\overline{\rho}$ = 20% with and without residual stress: (**a**) P-S, (**b**) G-S, (**c**) IWP-S, and (**d**) IWP-L.

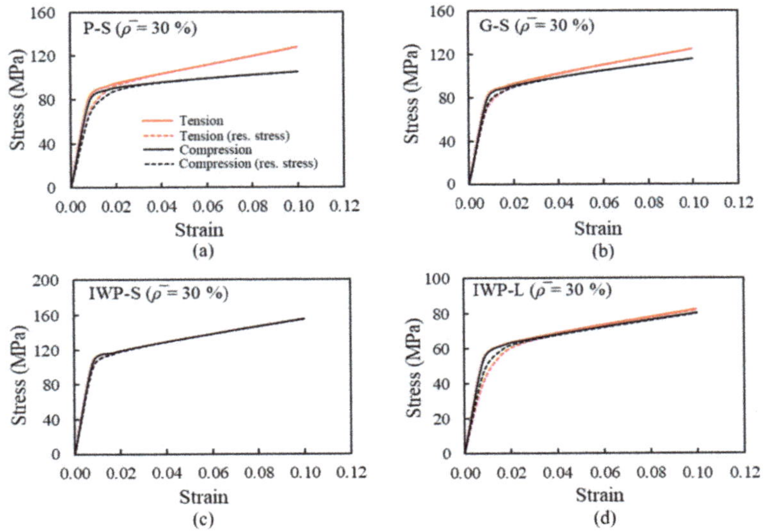

Figure 13. The stress–strain curve for tension/compression loading for $\overline{\rho}$ = 30% with and without residual stress: (**a**) P-S, (**b**) G-S, (**c**) IWP-S, and (**d**) IWP-L.

3.2. Effective Elastic Modulus and Yield Strength

The elastic modulus and yield strength of the different TPMS lattices for the three different densities considered are extracted from the stress–strain curves, for both tension and compression loading, with and without the residual stress effects. The Gibson–Ashby power law is used to correlate Young's modulus E and yield strength σ to relative densities $\overline{\rho}$ of TPMS structures and is given by the following equations [37];

$$E/E_0 = C_1\,(\overline{\rho})^n \tag{19}$$

$$\sigma/\sigma_0 = C_2\,(\overline{\rho})^m \tag{20}$$

where E_0 and σ_0 correspond to the Young's modulus and yield strength of the solid base material at room temperature (e.g., see Table 1), while E and σ are the elastic modulus and yield strength of the lattice at a relative density $\overline{\rho}$, C_1 and C_2 are the scaling coefficients, and n and m are the scaling exponents of the fitting curves.

The normalized Young's modulus (E/E_0) and normalized yield strength (σ/σ_0) values for different relative densities are shown in Figures 14 and 15, respectively. The elastic modulus and yield strength of the lattice structures increase with the increase in relative densities for all TPMS structures, as expected. Considering the effect of residual stresses, e.g., Figure 14, the

TPMS accounting for residual stresses shows a lower elastic modulus in the build *z*-direction compared with the reference TPMS without residual stresses. The effect of residual stress is most prominent on the sheet-based TPMS P-S in Figure 14a and the ligament-based IWP-L in Figure 14d. There is also a notable difference in stiffness under compression and tension loading. However, as seen in Figure 15, there is an insignificant effect of the residual stress on the yield strength in the build *z*-direction. Figures 16 and 17 show the amount of reduction (%) in Young's modulus due to the effect of residual stress in tensile and compressive loading, respectively. In tensile loading, the amount of reduction in elastic modulus due to residual stress is observed to be more prominent (>25%) in IWP-L for all relative densities, and for P-S structures, a reduction of more than 15% is observed at higher relative density, and the percentage reduction in elasticity was observed to be least in the case of IWP-S. For the compressive loading case, the reduction in elastic modulus is less prominent for IWP-L compared to in tensile loading and a very nominal reduction in elastic modulus is found in the case of G-S and IWP-S lattices due to residual stress. However, the P-S structure showed almost the same amount of reduction in elastic modulus compared to the tensile loading case. The power law coefficient and exponents of the scaling laws in Equations (19) and (20) for elastic modulus and yield strength for the four TPMS topologies are given in Table 3. The effect of the residual stress can be realized from the values of the fitting constant and exponent C_1 and n, respectively, on the elastic modulus, and similarly for the effect of indicating distinctly different values, especially for the IWP-L topology. The nature of elastic deformation behaviors of the TPMS structures under tensile/compression loading can be best explained using the exponent value n of the power law in Equation (19) [38]. From the n exponent values of the respective TPMS in Table 3, in uniaxial tensile and compression loading, the sheet-based P-S, G-S, and IWP-S structures show a mixed mode of deformation (e.g., $1 < n < 2$), although stretching may be more pronounced, while the ligament-based IWP-L lattice shows a predominantly bending behavior ($n \geq 2$), indicating that deformation behavior is strongly related to the topology of the TPMS. These observations agree well with the behavior reported in the literature on TPMSs [26,39]. In addition, it can be observed that due to predominantly bending behavior, the IWP-L shows less mechanical stiffness and strength compared to the sheet-based TPMS of the same relative densities as indicated by the respective values of magnitude in Figures 14 and 15.

Table 3. Scaling law parameters, normalized young's modulus, and yield strength.

P-S	Young's Modulus		Yield Strength		G-S	Young's Modulus		Yield Strength	
	C_1	n	C_2	m		C_1	n	C_2	m
Tension	0.34	1.45	0.76	1.33	Tension	0.29	1.33	0.68	1.29
Tension (residual stress)	0.26	1.38	0.72	1.30	Tension (residual stress)	0.22	1.25	0.67	1.29
Compression	0.34	1.47	0.74	1.36	Compression	0.29	1.35	0.60	1.20
Compression (residual stress)	0.28	1.46	0.70	1.33	Compression (residual stress)	0.27	1.32	0.60	1.20
Lee et al. [28]	0.61	1.57	0.79	1.36	Al-Ketan et al. [23]	0.51	1.38	0.44	1.24
IWP-S	Young's Modulus		Yield Strength		IWP-L	Young's Modulus		Yield Strength	
	C_1	n	C_2	m		C_1	n	C_2	m
Tension	0.36	1.27	0.76	1.20	Tension	0.72	2.41	1.26	2.13
Tension (residual stress)	0.33	1.25	0.77	1.20	Tension (residual stress)	0.51	2.40	1.27	2.13
Compression	0.38	1.33	0.75	1.20	Compression	0.72	2.43	1.27	2.13
Compression (residual stress)	0.36	1.32	0.76	1.20	Compression (residual stress)	0.61	2.41	1.23	2.11
Al-Ketan et al. [26]	0.79	1.48	0.69	1.24	Al-Ketan et al. [26]	1.28	2.64	1.05	2.14

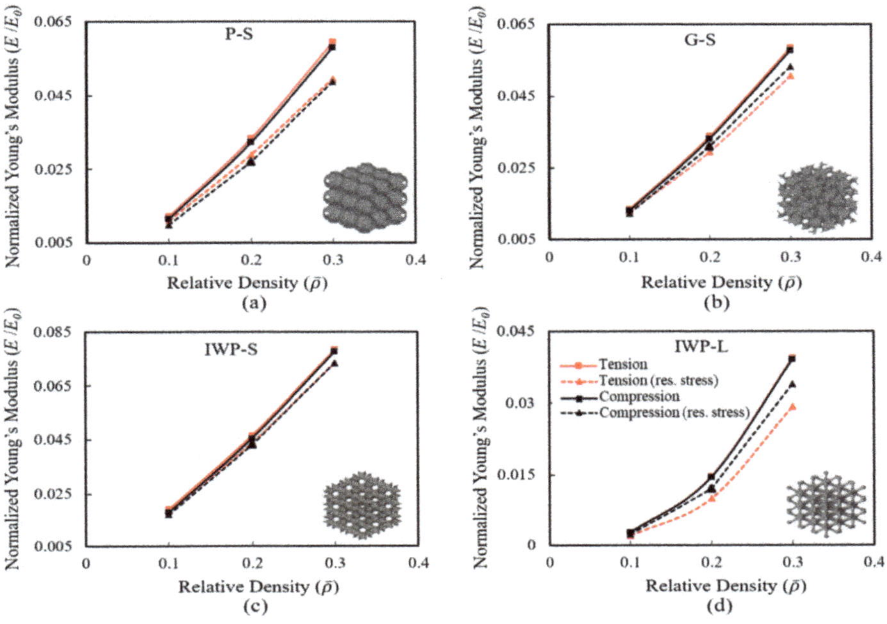

Figure 14. Effect of residual stresses on the normalized Young's modulus (E/E_0) with respect to relative density $\bar{\rho}$: (**a**) P-S, (**b**) G-S, (**c**) IWP-S, and (**d**) IWP-L.

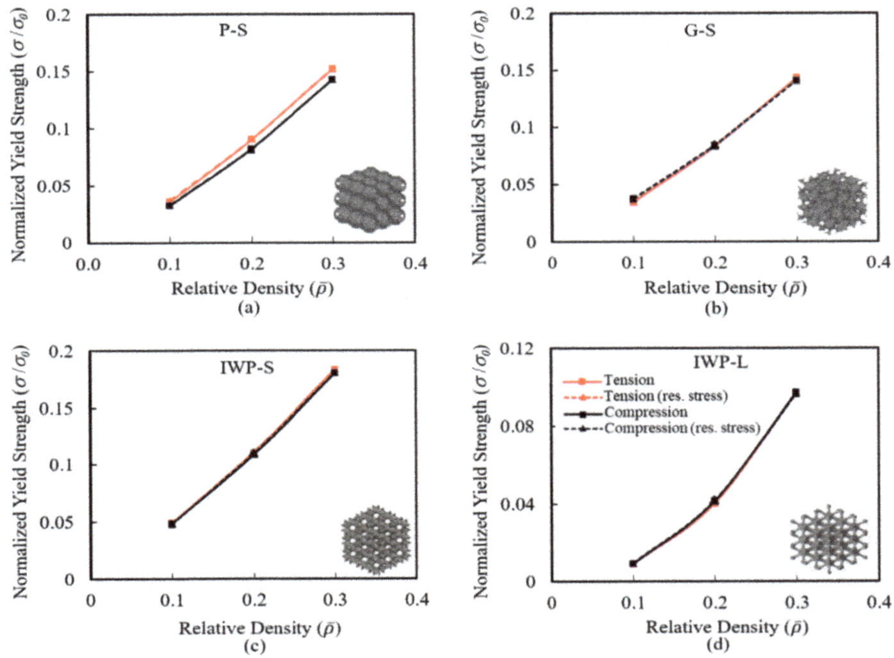

Figure 15. Effect of residual stresses on the normalized yield strength (σ/σ_0) with respect to relative density $\bar{\rho}$: (**a**) P-S, (**b**) G-S, (**c**) IWP-S, and (**d**) IWP-L.

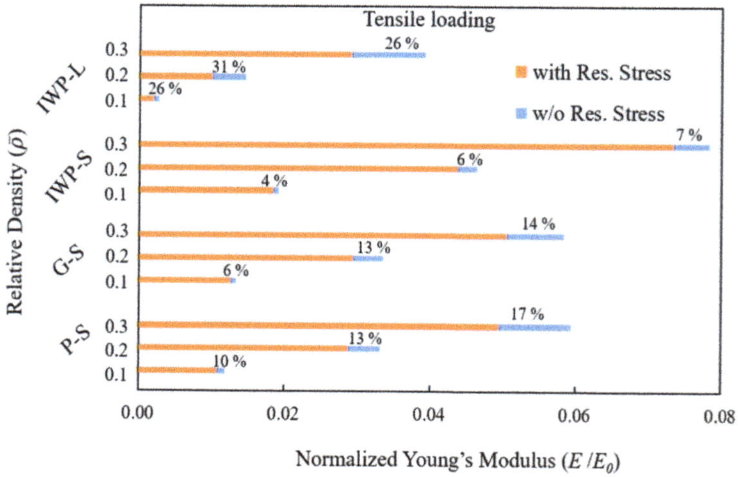

Figure 16. The reduction in Young's modulus in percentages for tensile loading for all TPMSs and their relative densities.

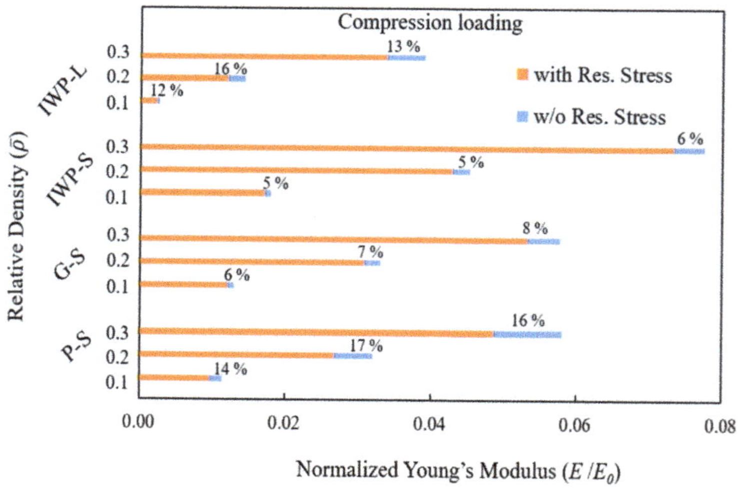

Figure 17. The reduction in Young's modulus in percentages for compressive loading for all TPMSs and their relative densities.

3.3. SEA Simulation

The specific energy absorption (SEA) value is defined as the energy absorbed by a material under uniaxial compression, normalized by its density, and involves large deformation up to the point of densification of the lattice structure. The objective here is to investigate the effect of the residual stress on the SEA capacity. The SEA is expressed as

$$\text{SEA} = \frac{1}{\bar{\rho}} \int_0^{\varepsilon_D} \sigma \, d\varepsilon \tag{21}$$

where σ is the compressive stress, ε is the compressive strain, and ε_D is the densification strain. The full-scale FEA models of the two TPMS configurations shown in Figure 18 are subjected to compressive loads to investigate the SEA value with and without accounting for residual stresses for different relative densities. The lower rigid plate is fixed, whereas the upper rigid plate is given a prescribed total displacement of Δ = 15 mm, corresponding to a total compressive strain of 50% in the build *z*-direction. A general contact interaction is defined between the upper and lower rigid plates, and the TPMS lattice with a frictional coefficient of 0.1 and the self-contact between surfaces of the TPMS are assigned a friction coefficient of 0.2.

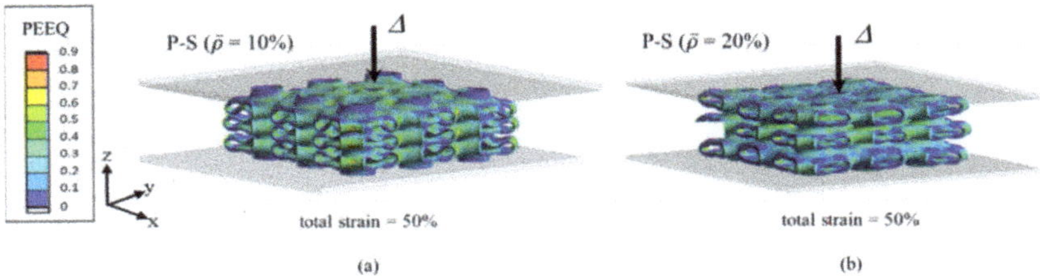

Figure 18. Equivalent plastic strain (PEEQ) contour plots at 50% deformation P-S with: (**a**) $\bar{\rho}$ = 10% and (**b**) $\bar{\rho}$ = 20%.

Figure 18 shows the equivalent plastic strain (PEEQ) for the P-S lattice for $\bar{\rho}$ = 10% and 20%. The reaction force on the upper rigid plate is obtained from which the effective compressive stress over the lattice is determined for both the cases with and without residual stresses accounted for. As shown in Figure 19, there is a change in elastic modulus as observed in previous sections due to residual stress; however, beyond the yield region, the stress–strain response reveals that the effect of the residual stresses is insignificant on the SEA capacity of the lattice structure, and is found to be same as compared to the reference TPMS, e.g., without residual stresses.

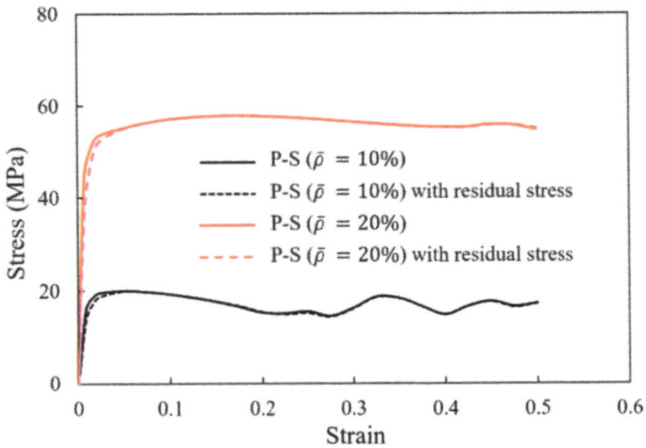

Figure 19. Stress–strain compression response for P-S lattice of $\bar{\rho}$ = 10% and 20%, with and without the effect of residuals tresses.

4. Conclusions

The effective mechanical properties such as elastic modulus, yield strength, and SEA of four TPMS structures with varying relative densities ($\bar{\rho}$ = 10%, 20% and 30%) are investigated considering the influence of residual stress evaluated from the SLM process through sequentially coupled thermomechanical simulations. When incorporating the residual stresses arising from the SLM process, the effective elastic modulus of all TPMS lattices in the build direction is lower in comparison with the reference TPMS in which the residual stresses are not accounted for. The decrease in stiffness is more prominent in sheet-based P-S and ligament-based IWP-L due to the higher accumulation of residual stress, and a stiffness reduction of more than 25% is observed in IWP-L. In addition, the fitting constant and exponent C_1 and n, respectively, in the Gibson–Ashby power law show distinctly different values, especially for the IWP-L topology, indicating the influence of residual stress on the elastic modulus of the TPMS. The sheet-based TPMS (P-S, G-S, and IWP-S) with a relative density of 10% show softening behavior beyond the yield region in compression loading, while for higher densities of all TPMSs, the tensile and compression behaviors show consistent hardening. The influence of residual stresses on yield strength and SEA is insignificant in the build direction for all TPMSs, indicating that residual stresses have little effect beyond the yield region in both compression and tension loading.

The findings with regard to the variation in stiffness due to residual stress inherited from the SLM process are significant as stiffness is known to considerably affect the vibration and fatigue properties of parts and hence needs to be accounted for in any design process. Further investigations for the anisotropic behavior of the TPMS structures considering the loading in other directions, the effect of various scanning strategies and post-heat treatment procedures on residual stress mitigations in TPMS lattices, and its influence on mechanical properties may be undertaken.

Author Contributions: Conceptualization, I.B. and R.K.A.A.-R.; methodology, N.A., I.B. and R.K.A.A.-R.; software, N.A.; validation, N.A.; formal analysis, N.A.; investigation, N.A.; resources, I.B. and R.K.A.A.-R.; data curation, N.A.; writing—original draft preparation, N.A.; writing—review and editing, I.B. and R.K.A.A.-R.; visualization, N.A.; supervision, I.B. and R.K.A.A.-R.; project administration, I.B. and R.K.A.A.-R.; funding acquisition, I.B. and R.K.A.A.-R. All authors have read and agreed to the published version of the manuscript.

Funding: The authors acknowledge the financial support provided by Khalifa University under Award No. RCII-2019-003.

Institutional Review Board Statement: Not applicable.

Informed Consent Statement: Not applicable.

Data Availability Statement: Not applicable.

Conflicts of Interest: The authors declare no conflict of interest.

Appendix A. Verification of Residual Stresses

For verification of residual stress values obtained through the simulations in the current study, experimentally measured values from the published literature [35] are used. The referenced work uses the neutron diffraction method to measure the inherited residual stress in a curved thin-walled plate fabricated through a metallic LPBF process. Figure A1a shows the geometric dimensions of the build made of Inconel 625 (IN625) material grade deposited on a stainless-steel base plate. The example case selected for comparison is deposited in a vertical orientation with a powder layer thickness of 30 μm. The simulation model uses an element layer height of 0.3 mm as per the recommended size range mentioned in Section 2.3, and a coarse mesh size of 2 mm is used for the baseplate. The location of the residual stress measurement path (e.g., T1) is at mid-height of the build indicated by the red dotted line shown in Figure A1a. Figure A1b depicts the comparison of stress component σ_z in the build direction along path T1, with markedly good agreement between the simulation results and the experimental measurements.

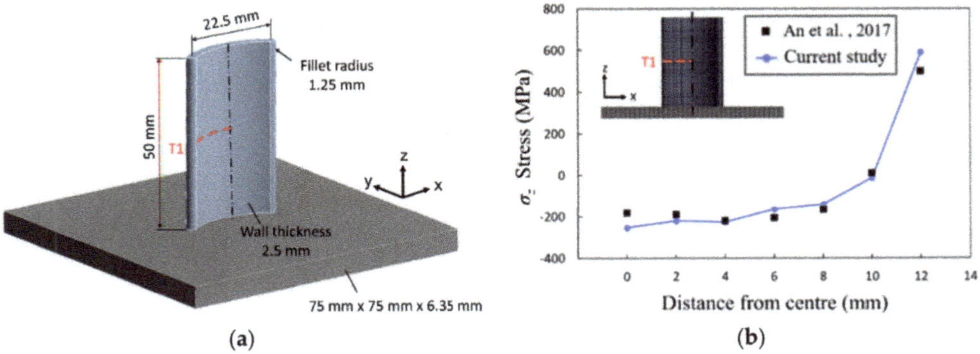

Figure A1. Sample used for comparison with current simulation study: (**a**) geometric dimension of the curve plate and stress measurement location (T1) indicated by red dotted line; (**b**) comparison of simulated residual stress component σ_z with experimental measurements from literature [35].

References

1. Hossain, U.; Ghouse, S.; Nai, K.; Rt, J. Controlling and Testing Anisotropy in Additively Manufactured Stochastic Structures. *Addit. Manuf.* **2021**, *39*, 101849. [CrossRef]
2. Maconachie, T.; Leary, M.; Lozanovski, B.; Zhang, X.; Qian, M.; Faruque, O.; Brandt, M. SLM Lattice Structures: Properties, Performance, Applications and Challenges. *Mater. Des.* **2019**, *183*, 108137. [CrossRef]
3. Zhang, L.; Song, B.; Fu, J.J.; Wei, S.S.; Yang, L.; Yan, C.Z.; Li, H.; Gao, L.; Shi, Y.S. Topology-Optimized Lattice Structures with Simultaneously High Stiffness and Light Weight Fabricated by Selective Laser Melting: Design, Manufacturing and Characterization. *J. Manuf. Process.* **2020**, *56*, 1166–1177. [CrossRef]
4. Al-Ketan, O.; Abu Al-Rub, R.K. Multifunctional Mechanical Metamaterials Based on Triply Periodic Minimal Surface Lattices. *Adv. Eng. Mater.* **2019**, *21*, 1–39. [CrossRef]
5. Maskery, I.; Aremu, A.O.; Parry, L.; Wildman, R.D.; Tuck, C.J.; Ashcroft, I.A. Effective Design and Simulation of Surface-Based Lattice Structures Featuring Volume Fraction and Cell Type Grading. *Mater. Des.* **2018**, *155*, 220–232. [CrossRef]
6. Ahmed, N.; Barsoum, I.; Haidemenopoulos, G.; Al-Rub, R.K.A. Process Parameter Selection and Optimization of Laser Powder Bed Fusion for 316L Stainless Steel: A Review. *J. Manuf. Process.* **2022**, *75*, 415–434. [CrossRef]
7. Liu, Y.; Yang, Y.; Wang, D. A Study on the Residual Stress during Selective Laser Melting (SLM) of Metallic Powder. *Int. J. Adv. Manuf. Technol.* **2016**, *87*, 647–656. [CrossRef]
8. Mercelis, P.; Kruth, J.P. Residual Stresses in Selective Laser Sintering and Selective Laser Melting. *Rapid Prototyp. J.* **2006**, *12*, 254–265. [CrossRef]
9. Parry, L.A.; Ashcroft, I.A.; Wildman, R.D. Geometrical Effects on Residual Stress in Selective Laser Melting. *Addit. Manuf.* **2019**, *25*, 166–175. [CrossRef]
10. Withers, P.J.; Bhadeshia, H.K.D.H. Residual Stress Part 2—Nature and Origins. *Mater. Sci. Technol.* **2001**, *17*, 366–375. [CrossRef]
11. Withers, P.J.; Bhadeshia, H.K.D.H. Residual Stress Part 1—Measurement Techniques. *Mater. Sci. Technol.* **2001**, *17*, 355–365. [CrossRef]
12. Rangaswamy, P.; Griffith, M.L.; Prime, M.B.; Holden, T.M.; Rogge, R.B.; Edwards, J.M.; Sebring, R.J. Residual Stresses in LENS® Components Using Neutron Diffraction and Contour Method. *Mater. Sci. Eng. A* **2005**, *399*, 72–83. [CrossRef]
13. Nytra, M.; Kubı, P. A Fully Coupled Thermomechanical Damage Analysis of Hot Closed Die Forging Using Finite Element Modeling. *J. Mater. Eng. Perform.* **2020**, *29*, 8236–8246. [CrossRef]
14. Denlinger, E.R.; Gouge, M.; Irwin, J.; Michaleris, P. Thermomechanical Model Development and in Situ Experimental Validation of the Laser Powder-Bed Fusion Process. *Addit. Manuf.* **2017**, *16*, 73–80. [CrossRef]
15. Mukherjee, T.; Zhang, W.; DebRoy, T. An Improved Prediction of Residual Stresses and Distortion in Additive Manufacturing. *Comput. Mater. Sci.* **2017**, *126*, 360–372. [CrossRef]
16. Ali, H.; Ghadbeigi, H.; Mumtaz, K. Effect of Scanning Strategies on Residual Stress and Mechanical Properties of Selective Laser Melted Ti6Al4V. *Mater. Sci. Eng. A* **2018**, *712*, 175–187. [CrossRef]
17. Schoinochoritis, B.; Chantzis, D.; Salonitis, K. Simulation of Metallic Powder Bed Additive Manufacturing Processes with the Finite Element Method: A Critical Review. *Proc. Inst. Mech. Eng. Part B J. Eng. Manuf.* **2017**, *231*, 96–117. [CrossRef]
18. Chen, Q.; Taylor, H.; Takezawa, A.; Liang, X.; Jimenez, X.; Wicker, R.; To, A.C. Island Scanning Pattern Optimization for Residual Deformation Mitigation in Laser Powder Bed Fusion via Sequential Inherent Strain Method and Sensitivity Analysis. *Addit. Manuf.* **2021**, *46*, 102116. [CrossRef]
19. Mirkoohi, E.; Sievers, D.E.; Garmestani, H.; Liang, S.Y. Thermo-Mechanical Modeling of Thermal Stress in Metal Additive Manufacturing Considering Elastoplastic Hardening. *CIRP J. Manuf. Sci. Technol.* **2020**, *28*, 52–67. [CrossRef]

20. Al-Ketan, O.; Lee, D.W.; Rowshan, R.; Abu Al-Rub, R.K. Functionally Graded and Multi-Morphology Sheet TPMS Lattices: Design, Manufacturing, and Mechanical Properties. *J. Mech. Behav. Biomed. Mater.* **2020**, *102*, 103520. [CrossRef]
21. Afshar, M.; Pourkamali Anaraki, A.; Montazerian, H. Compressive Characteristics of Radially Graded Porosity Scaffolds Architectured with Minimal Surfaces. *Mater. Sci. Eng. C* **2018**, *92*, 254–267. [CrossRef] [PubMed]
22. Montazerian, H.; Davoodi, E.; Asadi-Eydivand, M.; Kadkhodapour, J.; Solati-Hashjin, M. Porous Scaffold Internal Architecture Design Based on Minimal Surfaces: A Compromise between Permeability and Elastic Properties. *Mater. Des.* **2017**, *126*, 98–114. [CrossRef]
23. Al-Ketan, O.; Rezgui, R.; Rowshan, R.; Du, H.; Fang, N.X.; Abu Al-Rub, R.K. Microarchitected Stretching-Dominated Mechanical Metamaterials with Minimal Surface Topologies. *Adv. Eng. Mater.* **2018**, *20*, 1800029. [CrossRef]
24. Dalaq, A.S.; Abueidda, D.W.; Abu Al-Rub, R.K.; Jasiuk, I.M. Finite Element Prediction of Effective Elastic Properties of Interpenetrating Phase Composites with Architectured 3D Sheet Reinforcements. *Int. J. Solids Struct.* **2016**, *83*, 169–182. [CrossRef]
25. Abu Al-Rub, R.K.; Lee, D.-W.; Khan, K.A.; Palazotto, A.N. Effective Anisotropic Elastic and Plastic Yield Properties of Periodic Foams Derived from Triply Periodic Schoen's I-WP Minimal Surface. *J. Eng. Mech.* **2020**, *146*, 04020030. [CrossRef]
26. Al-Ketan, O.; Abu Al-Rub, R.K.; Rowshan, R. The Effect of Architecture on the Mechanical Properties of Cellular Structures Based on the IWP Minimal Surface. *J. Mater. Res.* **2018**, *33*, 343–359. [CrossRef]
27. Yang, E.; Leary, M.; Lozanovski, B.; Downing, D.; Mazur, M.; Sarker, A.; Khorasani, A.M.; Jones, A.; Maconachie, T.; Bateman, S.; et al. Effect of Geometry on the Mechanical Properties of Ti-6Al-4V Gyroid Structures Fabricated via SLM: A Numerical Study. *Mater. Des.* **2019**, *184*, 108165. [CrossRef]
28. Lee, D.W.; Khan, K.A.; Abu Al-Rub, R.K. Stiffness and Yield Strength of Architectured Foams Based on the Schwarz Primitive Triply Periodic Minimal Surface. *Int. J. Plast.* **2017**, *95*, 1–20. [CrossRef]
29. Zheng, X.; Fu, Z.; Du, K.; Wang, C.; Yi, Y. Minimal Surface Designs for Porous Materials: From Microstructures to Mechanical Properties. *J. Mater. Sci.* **2018**, *53*, 10194–10208. [CrossRef]
30. Abueidda, D.W.; Abu Al-Rub, R.K.; Dalaq, A.S.; Lee, D.W.; Khan, K.A.; Jasiuk, I. Effective Conductivities and Elastic Moduli of Novel Foams with Triply Periodic Minimal Surfaces. *Mech. Mater.* **2016**, *95*, 102–115. [CrossRef]
31. Maskery, I.; Sturm, L.; Aremu, A.O.; Panesar, A.; Williams, C.B.; Tuck, C.J.; Wildman, R.D.; Ashcroft, I.A.; Hague, R.J.M. Insights into the Mechanical Properties of Several Triply Periodic Minimal Surface Lattice Structures Made by Polymer Additive Manufacturing. *Polymer* **2018**, *152*, 62–71. [CrossRef]
32. Waqar, S.; Guo, K.; Sun, J. FEM Analysis of Thermal and Residual Stress Profile in Selective Laser Melting of 316L Stainless Steel. *J. Manuf. Process.* **2021**, *66*, 81–100. [CrossRef]
33. *ANSYS Analysis User's Manual Version*; ANSYS: Canonsburg, PA, USA, 2020.
34. Al-Ketan, O.; Abu Al-Rub, R.K. MSLattice: A free software for generating uniform and graded lattices based on triply periodic minimal surfaces. *Mater. Des. Process. Commun.* **2020**, *3*, e205. [CrossRef]
35. An, K.; Yuan, L.; Dial, L.; Spinelli, I.; Stoica, A.D.; Gao, Y. Neutron Residual Stress Measurement and Numerical Modeling in a Curved Thin-Walled Structure by Laser Powder Bed Fusion Additive Manufacturing. *Mater. Des.* **2017**, *135*, 122–132. [CrossRef]
36. Manvatkar, V.; De, A.; Debroy, T. Heat Transfer and Material Flow during Laser Assisted Multi-Layer Additive Manufacturing. *J. Appl. Phys.* **2014**, *116*, 1–8. [CrossRef]
37. Gibson, L.J.; Ashby, M.F. *Cellular Solids: Structure and Properties*, 1st ed.; Cambridge University Press: Cambridge, UK, 2001.
38. Ejeh, C.J.; Barsoum, I.; Abu Al-Rub, R.K. Flexural Properties of Functionally Graded Additively Manufactured AlSi10Mg TPMS Latticed-Beams. *Int. J. Mech. Sci.* **2022**, *223*, 107293. [CrossRef]
39. Al-Ketan, O.; Rowshan, R.; Abu Al-Rub, R.K. Topology-Mechanical Property Relationship of 3D Printed Strut, Skeletal, and Sheet Based Periodic Metallic Cellular Materials. *Addit. Manuf.* **2018**, *19*, 167–183. [CrossRef]

metals

Article

Influence of Powder Characteristics on the Microstructure and Mechanical Behaviour of GH4099 Superalloy Fabricated by Electron Beam Melting

Shixing Wang [1,2], Shen Tao [3] and Hui Peng [3,4,*]

[1] School of Materials Science and Engineering, Beihang University, 37 Xueyuan Road, Beijing 100191, China; wsx161@163.com
[2] Surface Engineering Research Institute, Chinese Academy of Agricultural Mechanization Sciences, No. 1 Beishatan Road, Beijing 100191, China
[3] Research Institute for Frontier Science, Beihang University, 37 Xueyuan Road, Beijing 100191, China; taoshen@buaa.edu.cn
[4] Key Laboratory of High-Temperature Structural Materials & Coatings Technology, Ministry of Industry and Information Technology, Beihang University, 37 Xueyuan Road, Beijing 100191, China
* Correspondence: penghui@buaa.edu.cn; Tel.: +86-(010)-82317117

Abstract: A Chinese superalloy, GH4099 (~20 vol.% γ' phase), which can operate for long periods of time at temperatures of 1173–1273 K, was fabricated by electron beam melting (EBM). Argon gas atomized (GA) and plasma rotation electrode process (PREP) powders with similar composition and size distribution were used as raw materials for comparison. The microstructure and mechanical properties of both the as-EBMed and post-treated alloy samples were investigated. The results show that the different powder characteristics result in different build temperatures for GA and PREP samples, which are 1253 K and 1373 K, respectively. By increasing the building temperature, the EBM processing window shifts towards a higher scanning speed direction. Microstructure analysis reveals that both as-EBM samples show a similar grain width (measured to be ~200 μm), while the size of γ' precipitated in the PREP sample (~90 nm) is larger than that of the GA sample (~130 nm) due to the higher build temperature. Fine spherical γ' phase precipitates uniformly after heat treatment (HT). Furthermore, intergranular cracking was observed for the as-fabricated PREP sample as a result of local enrichment of Si at grain boundaries. The cracks were completely eliminated by hot isostatic pressing (HIP) and did not re-open during subsequent heat treatment (HT) of solution treatment and aging. The tensile strength of the PREP sample after HIP and HT is ~920 MPa in the building direction and ~850 MPa in the horizontal direction, comparable with that of the wrought alloy.

Keywords: Ni-based superalloys; electron beam melting; additive manufacturing; Argon gas atomized; plasma rotation electrode process; powder characteristics

Citation: Wang, S.; Tao, S.; Peng, H. Influence of Powder Characteristics on the Microstructure and Mechanical Behaviour of GH4099 Superalloy Fabricated by Electron Beam Melting. *Metals* **2022**, *12*, 1301. https://doi.org/10.3390/met12081301

Academic Editor: Zhanyong Zhao

Received: 7 July 2022
Accepted: 1 August 2022
Published: 3 August 2022

1. Introduction

Additive manufacturing (AM) is one of the major driving innovations in different industries. Typically, AM is a powerful tool for the aerospace industry [1–3]. Powder bed fusion (PBF) additive manufacturing, including both selective laser melting (SLM) and electron beam melting (EBM), does not require structural supports, which can optimize the AM building process. In addition, PBF techniques offer higher precision compared with other AM techniques [3]. Recent literature demonstrates that several materials were successfully fabricated by SLM, including 17-4 Ph steel [4], 316 L stainless steel [5], and Ti6Al4V [6,7]. Compared with SLM, the major feature of EBM is the preheated powder bed, which helps alleviate thermal stress during rapid solidification.

Ni-based superalloys are widely used in hot sections in the aerospace industry for their excellent high temperature mechanical properties and oxidation/corrosion resistance [8].

Both SLM and EBM have been utilized for fabricating complex superalloy parts [9,10]. Several previous works have demonstrated the feasibility of fabricating dense and crack free superalloys with a tailored microstructure [11–16]. An ultra-fine crystallographic grain structure was commonly observed in the SLMed superalloys, owing to the rapid solidification rate and high temperature gradients within the micro melt pool. By contrast, coarser columnar grains were usually formed in the EBMed superalloys under preheated conditions [9,17]. Particularly, by carefully controlling the process parameters, a columnar to equiaxed (CET) transition of grains was realized [18–21]. It has also been demonstrated that the mechanical properties of some AM superalloys are comparable to those of wrought and cast superalloys [22,23].

According to the content of Al and Ti, superalloys can be divided into two categories, weldable or nonweldable [12,13,24]. The level of Al + Ti determines the volume fraction of γ' precipitates within the superalloy, which plays an important role in the high-temperature strength and creep resistance [23,25,26]. Unfortunately, superalloys with high amount of γ' precipitates are considered to be difficult to weld because of their high tendency to crack, particularly as a result of the phenomenon of hot cracking [11,27]. Current SLM and EBM are developed based on welding technologies. However, the processing of nonweldable superalloys brings challenges to AM.

Some recent published works have focused on the cracking mechanisms of AM nonweldable superalloys. It was revealed in Ref. [27] that the hot cracking behavior is attributed to the presence of a liquid film during the last stage of solidification and thermal stresses. A correlation between cracks and high angle grain boundaries (HAGB) was identified. It was also confirmed by atom probe tomography (APT) that a local enrichment of minor element B at the grain boundaries was responsible for the formation of a liquid film at HAGB [27]. Further works performed by Körner [23] and Martin [15] demonstrated the feasibility of producing single crystal superalloy by EBM, in which crack free samples of a nonweldable Ni-based superalloy were successfully obtained by eliminating grain boundaries via competitive grain growth. A very recent work proposed the atomic-grain boundary design [11]. Although these innovative results have aroused widespread interest in the additive manufacturing of superalloys, hot cracks yielded in polycrystalline nonweldable superalloys (columnar or equiaxed grain structure) remain unresolved.

Decreasing the level of γ' precipitates in the nonweldable superalloy can reduce the cracking tendency significantly. However, a proper amount of γ' precipitates is desirable for adequate mechanical strength. For certain high-temperature applications, the amount of γ' precipitates should be optimized to achieve the best balance of strength and weldability. An example of this is the relatively newly designed γ'-strengthened superalloy, HAYNES 282, which was developed for applications in both aero and land-based gas turbine engines [28]. The total γ' forming elements (Al + Ti) content of HAYNES 282 is 3.7 wt.%, with the γ' mole fraction of ~19%. The carefully selected γ' precipitates level ensures the alloy has a combination of satisfied mechanical properties and improved weldability, compared with the typical non weldable superalloy.

In this work, a Chinese superalloy called GH4099 was fabricated by electron beam melting. This alloy has a similar γ' precipitates level with HAYNES 282, and can operate for a long term at temperatures of 1173–1273 K. The objective of the following investigation is to exhibit the influence of raw powder characteristics on the microstructure of the EBMed superalloy. The mechanism of minor element Si on the cracking behavior was elucidated. The evolution of cracks during heat treatment and mechanical properties of the samples were also of concern.

2. Experiment

2.1. GH4099 Powders

Argon gas atomized (GA) and plasma rotating electrode process (PREP) GH4099 powders were used as raw material, which were provided by CASIC-Hunan and CISRI-Gaona, China, respectively. The chemical composition of both GA and PREP powders were measured by inductively coupled plasma atomic emission spectroscopy (ICP-OES), as shown in Table 1. The nominal composition of GH4099 specified by *China Aeronautical Materials Handbook* is also listed in Table 2 for comparison. It can be seen that the two powders were provided with similar composition, except for the fact that the Si content in the PREP powders is almost one order of magnitude higher than that of the GA powder. Note that the 0.285 wt.% of Si is quite close to the permissible limit of the composition standard. Powder morphologies are shown in Figure 1. Both of the powders are spherical. However, the presence of small satellites and micro particles can be observed for the GA powders, compared with the perfectly smooth PREP powders. Figure 2 exhibits similar size distribution curves for the two powders measured by a laser granulometry (Bettersize, 9300ST). The powder size ranges primarily between 45 μm and 140 μm, with an average value of ~90 μm. The tap density of two powders was determined to be 4.53 g/cm^3 and 4.89 g/cm^3, respectively. Flow times of 16.4 s and 15.4 s were measured by a Hall flowmeter (50 g, 2.54 mm orifice), see Table 3.

Table 1. Chemical composition of GA and PREP powders of GH4099 (wt.%).

Powder Type	**Element Content, wt.%**							
	Cr	**Co**	**W**	**Mo**	**Ti**	**Al**	**Fe**	**Mn**
Argon Gas Atomized (GA)	18.21	6.44	5.88	4.00	1.29	2.03	<0.1	<0.005
	Si	C	N	O	Ni			
	0.025	<0.1	0.005	0.008	Bal.			
Plasma Rotating Electrode Process (PREP)	Cr	Co	W	Mo	Ti	Al	Fe	Mn
	18.65	5.91	5.91	3.89	1.29	2.26	0.54	0.24
	Si	C	N	O	Ni			
	0.285	<0.1	0.003	0.008	Bal.			

Table 2. Nominal composition of GH4099 (wt.%).

Element	**Ni**	**Mo**	**Fe**	**Cr**	**W**	**Al**	**Ti**
Content	Bal.	3.50–5.00	≤3.00	17.50–19.50	5.50–7.00	2.50–3.00	1.00–1.50
Element	Co	Si	B	Mn	Cu	Ce	S/P
Content	5.00–8.00	≤0.30	≤0.005	≤0.30	≤0.070	≤0.020	≤0.015

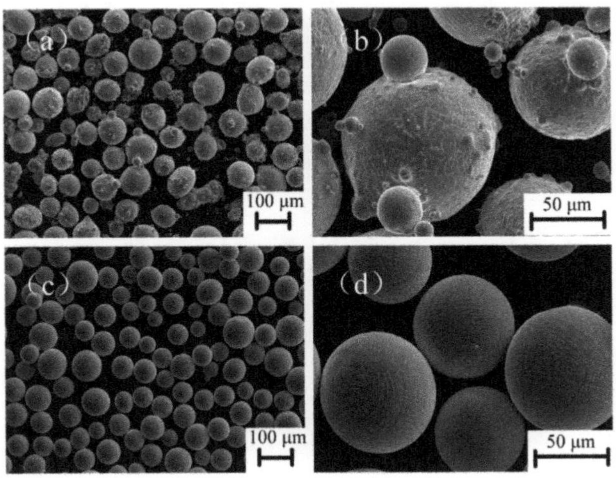

Figure 1. SEM morphology of pre-alloyed GH4099: (**a**,**b**) GA; (**c**,**d**) PREP.

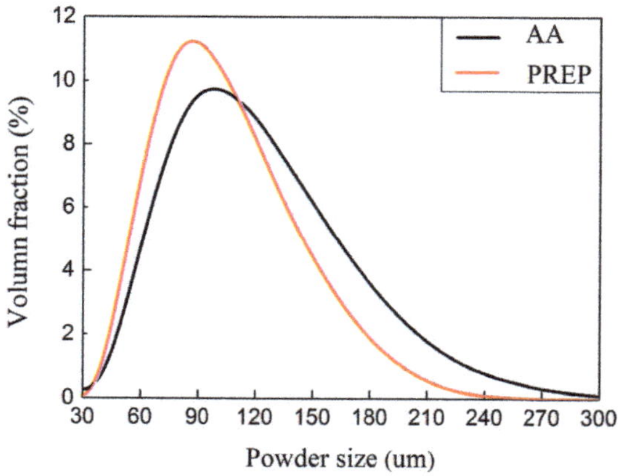

Figure 2. Volume fraction distribution of powder particles.

Table 3. Powder properties of two types of GH4099.

Powder	SSA (m^{-2}/kg)	Flowability (s)	Packing Density (g/cm^3)	D10 (μm)	D50 (μm)	D90 (μm)
GA	32.91	16.4	4.53	49.68	92.57	155.0
PREP	28.36	15.4	4.89	52.74	84.00	133.8

2.2. EBM Sample Fabrication

GH4099 superalloy samples were fabricated using an Arcam A2XX machine (Control Software 3.2, Arcam, Stockholm, Sweden) in a manual mode. Figure 3a shows the schematic diagram of one batch of as-EBM samples. Layers were scanned using a snake-like melting strategy (i.e., either 0° or 90° hatch direction was used for each layer), as shown in Figure 3b. A 10 mm thick stainless steel plate with a diameter of 150 mm was used as a starting plate. No supports were applied for the EBM fabrication. One batch of the EBM fabrication included 9 with each 20 value × 20 value × 30 value mm in X-Y-Z dimension. EBM process windows were first established for the GA powders by using a range of scanning speed v from 333–9000 mm/s, combined with a range of line offset L_{off} from 0.1–0.3 mm, followed by the PREP powders. Three different beam current I values of 5 mA, 10 mA, and 15 mA were selected. This resulted in equivalent area energy E_A values ranging from 0.414 J/mm^2 to 3 J/mm^2. This resulted in equivalent area energy E_A ($E_A = UI/(v{\cdot}L_{off})$, where U is the operating voltage of 60 kV) values ranging from 0.414 J/mm^2 to 3 J/mm^2. A similar EBM process window diagram was also established for the PREP powders. Detailed process parameters are summarized in Table 4.

Table 4. Summary of the processing parameters of two types of GH4099 powders.

GA Powder				PREP Powder			
I (mA)	v (mm/s)	L_{off} (mm)	E_A (J/mm^2)	I (mA)	v (mm/s)	L_{off} (mm)	E_A (J/mm^2)
5	333	0.3	3.00	5	3250	0.1	0.92
5	500	0.2	3.00	5	4250	0.1	0.71
5	833	0.3	1.20	5	3750	0.1	0.80
5	1000	0.1	3.00	10	417	0.3	4.80
5	1666	0.15	1.20	10	625	0.2	4.80
5	3000	0.1	1.00	10	833	0.3	2.40
5	4500	0.1	0.67	10	1250	0.1	4.80

Table 4. *Cont.*

GA Powder				PREP Powder			
I (mA)	v (mm/s)	L_{off} (mm)	E_A (J/mm^2)	I (mA)	v (mm/s)	L_{off} (mm)	E_A (J/mm^2)
10	833	0.3	2.40	10	1250	0.2	2.40
10	1250	0.2	2.40	10	1250	0.3	1.60
10	1250	0.3	1.60	10	1875	0.2	1.60
10	2000	0.1	3.00	10	2250	0.1	2.67
10	2500	0.1	2.40	10	2500	0.1	2.40
10	3000	0.1	2.00	10	3750	0.1	1.60
10	3500	0.1	1.71	10	4500	0.1	1.33
10	4000	0.1	1.50	10	6500	0.1	0.92
10	4500	0.1	1.33	10	8500	0.1	0.71
10	5000	0.1	1.20	10	9750	0.1	0.62
10	5500	0.1	1.09	10	10,500	0.1	0.57
10	6000	0.1	1.00	10	12,500	0.1	0.48
10	6500	0.1	0.92	10	14,500	0.1	0.41
15	1875	0.2	2.40	10	16,500	0.1	0.36
15	1500	0.1	6.00	10	18,500	0.1	0.32
15	2000	0.1	4.50	15	6750	0.1	1.33
15	2500	0.1	3.60	15	9750	0.1	0.92
15	3000	0.1	3.00	15	12,750	0.1	0.71
15	7500	0.1	1.20	15	2500	0.1	3.60
15	9000	0.1	1.00				
15	9000	0.1	1.00				

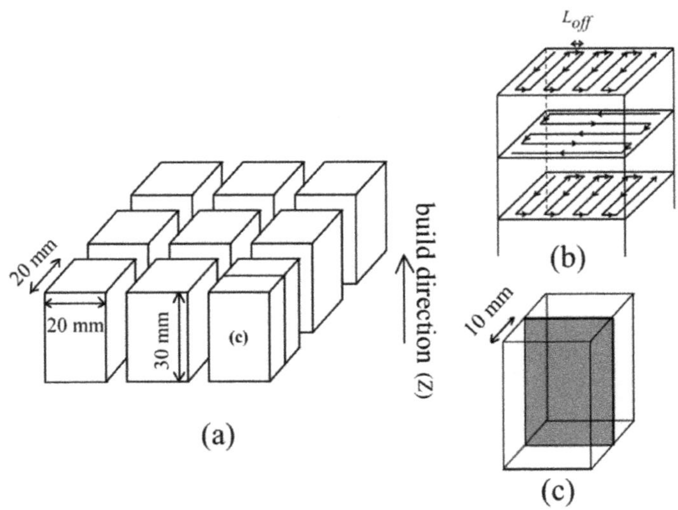

Figure 3. Schematic diagrams of (**a**) one batch of as-EBM samples; (**b**) electron beam scans in a cross snake-like way with a specific L_{off} and hatching layer of 50 μm; (**c**) shows the microstructural characterization region.

The build temperature for the GA and PREP powders was determined to be 1253 K and 1373 K, respectively, at which the powder bed was slightly sintered. The monitored temperature was almost constant for the whole EBM fabrication process. The samples were cooled slowly to room temperature after building.

2.3. Microstructural Characterisation

As-EBM samples were sectioned along the vertical plane, as illustrated in Figure 3c. Standard metallographic procedures including grinding and polishing were performed. A Gemini SEM 300 (Zeiss, Oberkochen, Germany) was used for scanning electron microscopy (SEM) for microstructural observation. Chemical etchant for revealing the grain boundaries was a solution of 4 g $CuSO_4$, 100 mL HCl, and 100 mL H_2O. The polished cross-sections were electro-etched at 5 V for 6 s in a solution of 12 mL H_3PO_4, 40 mL HNO_3, and 48 mL H_2SO_4 to examine the γ' precipitations. An electron-probe microanalyzer (EPMA) JXA-8100 (JEOL, Tokyo, Japan) was also employed to detect the minor element concentration at the vicinity of a crack. One thing that should be mentioned is that in order to prevent element contamination during sample preparation, the EPMA samples were cut by diamond wire, ground by diamond abrasive papers, and polished by diamond slurry.

2.4. Post-Processing

The samples with internal cracks were healed by hot isostatic pressing (HIP, Model QIH9, Quintus) at 1423 K, 150 MPa for 4 h, referred to as EBM+HIP condition. Selected EBM+HIP samples were then subjected to a solution and aging heat treatment (HT). The standard two-step treatment included 1453 K/2 h/AC (air cool) +1173 K/4 h/AC. This condition is referred to as EBM+HIP+HT. The HT was performed in a horizontal tube furnace in an air atmosphere. Throughout the HT process, a thermocouple was placed next to the samples to monitor the temperature. It is worth noting that the purpose of solution treatment is to make unevenly distributed γ' precipitations further re-dissolved into the γ' matrix, i.e., the uniform precipitation of γ' phases, and to readjust the morphology of γ' phase by changing aging temperature, aging time, or cooling method. This is done to realize precipitation strengthening and improve the mechanical properties of the Ni-based superalloy.

2.5. Mechanical Property Characterization

Micro-hardness measurements were performed using an FM-800 micro-hardness tester (FUTURE-TECH, Tokyo, Japan). A 200 g load force with a holding time of 10 s was used. Fifteen measurement points were made per sample on the sectioned plane (Figure 3c). The average hardness value is reported together with the standard deviation (STDEV).

Flat tensile specimens were machined with the load axes parallel and perpendicular to the build direction. Specimens in both as-EBM and post-processed conditions were prepared. Tensile tests were performed at room temperature using an electronic universal testing machine CMT4000 (SANS, Shenzhen, China), where strain was determined with an extensometer. A constant cross-head speed of 0.015 mm/s was used.

3. Results

3.1. Characteristics of Starting Powders for EBM Process

Gas atomization is completed by molten metal which is atomized due to inert gas jets into fine metal droplets, then cooled down in the atomizing tower. The plasma-rotating electrode process is a centrifugal atomization process in a vacuum atmosphere developed by Starmet. Plasma arc is involved in this process, and this method is currently a leading candidate for high-purity powders of metals with a high melting point such as Ti and Ni alloy powder production [29–31].

Representative micrographs (SEM) of atomized powders are shown in Figure 1a,b for GA powders and Figure 1c,d for PREP powders. In general, PREP powders shown in Figure 1c exhibited a much better spherical shape compared to GA powders, Figure 1a.

In addition, small satellite particles can be found in PREP powders, Figure 1c,d. As described in [32], powder particle size distribution and its mean size have traditionally been considered to be important. This has a certain impact on packing density on the build plate as well as the density of the finished part. Powders with smaller average particles (D50) have been found to result in better surface roughness of the completed samples according to [33]. Large particles might not be fully melted and fused, conversely too-small particles can lead to ball, spatter, or swell causing too much energy [34]. However, there are many particle properties that could potentially influence the overall behavior of the powder for the subsequent EBM process. These key particle properties include specific surface area, flowability, packing density, and D90.

Figure 4 shows the interior morphology of GH4099 powder particles. It was clear that the interior of GA powder particles were massive cellular dendrites (Figure 4a,b) due to rapid solidification during the cooling process. The interior morphology of PREPed GH4099 powder particles were mainly segregated dendritic and light from cellular dendrites (Figure 4c,d), resulting from different cooling rates between metal, liquid, and matrix in the solidification process [31], which made the dendrites grow and become coarse.

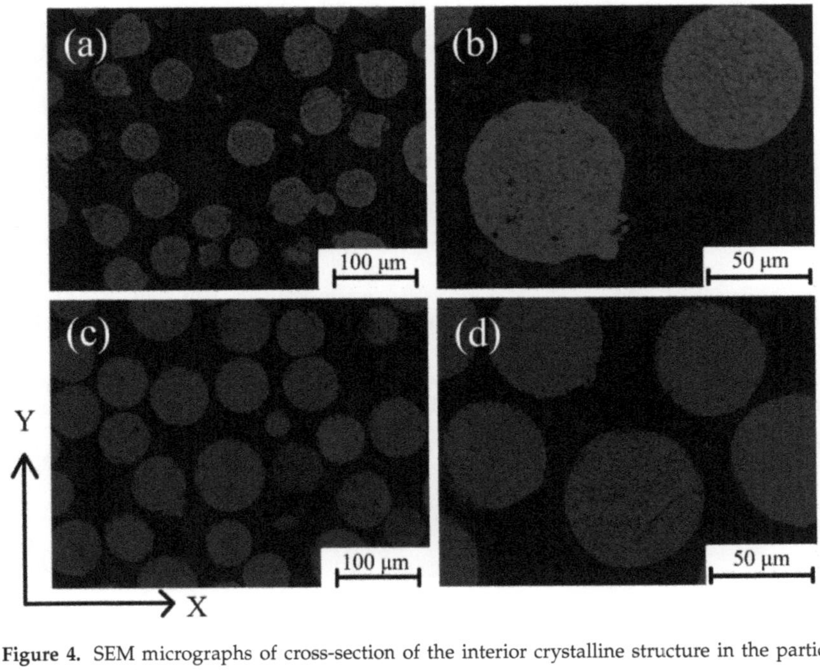

Figure 4. SEM micrographs of cross-section of the interior crystalline structure in the particles (**a**,**b**) GA; (**c**,**d**) PREP.

3.2. Process Window

According to the results of optical microscope (OM) and SEM, these samples have two types of defects. Namely uneven and bonding defects in the micromorphology occurred during the EBM process window based on the varying melting current and scanning speed. There were also dense samples with no surface unevenness, which were divided into two forming windows according to the different powders used in this work, as shown in Figure 5.

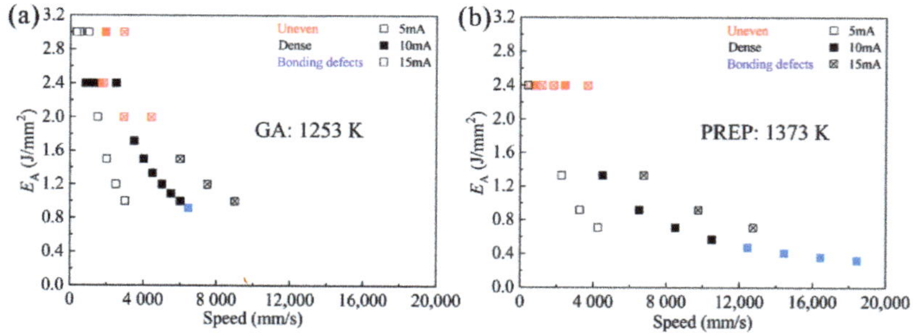

Figure 5. EBM processing windows for (**a**) GA powder and (**b**) PREP powder are plotted with scanning speed (v) against E_A. Suitable parameters (combinations of beam current, scanning speed, and E_A) for dense samples with no top surface unevenness and binding defects are indicated by black symbols.

From Figure 5a, the GA powders with the build temperature of 1253 K, unevenness or bonding defects occur when E_A is greater than ~2.0 J·m^{-2} or less than ~1.0 J·m^{-2}. For the PREP powders with the build temperature of 1373 K, dense samples could be obtained when E_A is in the range of ~0.5 J·m^{-2} to ~1.5 J·m^{-2} (Figure 5b). It can be concluded for both cases that a suitable combinations of parameters (beam current, scanning speed, and E_A) are necessary for defect free samples. By increasing the build temperature, the EBM processing window shifts towards higher speed and lower energy exposure per area E_A (in J·m^{-2}) direction. Due to the different powder characteristics, the PREP powders are more difficult to sinter during the preheating of the powder bed by electron beam. This resulted in a higher build temperature for the PREP powders than the GA powders.

Bonding defects occur with increased scanning speed when the energy input is too low to completely melt the powder layer, leaving the longitudinal pores [14]. Unevenness on the surface occurs with lower deflection speeds and a higher scanning current of the electron beam. Helmer et al. [35] reported that unevenness of the top surface is a result of a vigorous melt pool motion caused by high forces exerted on the melt pool which could be interpreted as temperature gradients, surface energy effects (Marangoni convection), and evaporation.

3.3. Microstructural Characterization

Columnar crystals and the equiaxial crystals were observed in as-EBM condition samples deposited with GA powder and PREP powder. An obvious columnar to equiaxed transition (CET) was also observed for both powders within the processing window, which is not shown here.

Two samples made from GA and PREP powders were selected for detailed investigations. As can be seen in Figure 6a,b, both samples reveal a similar columnar grain width of ~200 μm. The average grain width of the EBM sample deposited with GA powder (GA sample), as well as the sample as-EBM condition deposited with PREP powder (PREP sample) are 195.16 ± 18.46 μm and 206.47 ± 28.74 μm, respectively. SEM images of the γ' precipitations are shown in Figure 6c,d. The distribution of the γ' is quite homogeneous in the as-EBM state, with a volume fraction of about 20% for both samples. The γ' is seen to be spherical and very fine in size. Differing from the limited formation of γ' in the as-SLMed superalloys [36], the processing conditions (rapid solidification and high build temperature) of EBM fabrication may provide a proper thermal history for the precipitation of γ'. It is therefore not surprising that the higher build temperature resulted in a larger γ' precipitated (~90 nm) in the PREP sample than that of the GA sample (~130 nm).

Figure 6. Optical micrographs in the X-Z plane of GH4099 samples fabricated by EBM deposited with: (**a**) GA; (**c**) PREP powder, and SEM micrographs of γ' phase as-EBM GH4099 samples showing the morphology of γ' precipitates: (**b**) GA; (**d**)PREP.

3.4. Cracking in as-EBM GH4099 Superalloy

Macro cracks were not observed on the external surface and the wire-cut surface of the as-EBM GH4099 samples. However, the micro-cracks existed in the samples deposited with PREP powder by optical microscope and scanning electron microscope after grinding and polishing, but not in the samples deposited with GA powder. The cracks are parallel to the building direction, and no transverse cracks or lack of fusion appear, Figure 7. These microcracks are distributed along the columnar grain boundaries, which indicates that the microcracks are intergranular cracks.

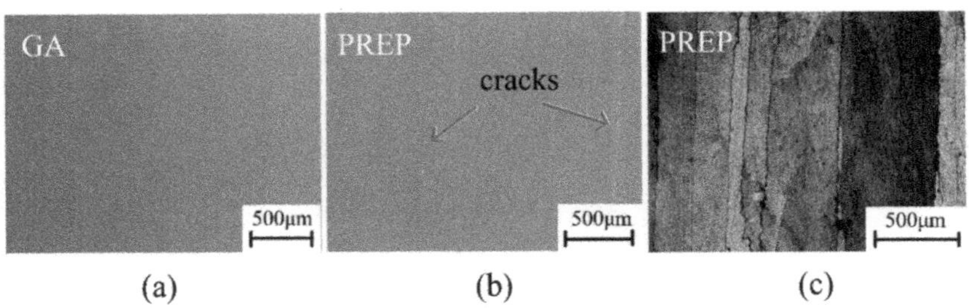

Figure 7. Crack morphology of as-EBM GH4099: optical micrographs of un-etched (**a**) GA sample with no cracks and (**b**) PREP sample with cracks parallel to the building direction. (**c**) Etched PREP sample showing the microcracks are intergranular cracks.

Observation of the polished cross-section are shown in Figure 7a,b. It can be seen that the GA sample is dense and crack-free. However, micro-cracks were detected in the PREP sample, with the cracking ratio determined to be ~5% in area fraction. The occurrence of cracks is associated with the formation of liquid films at grain boundaries, as indicated by the arrows shown in Figure 7b. Figure 7c also confirms that the microcracks are intergranular cracks.

3.5. Crack Healing and Microstructural Evolution

Kirka et al. [37] found that holes in Inconel 718 alloy formed by EBM could be removed by hot isostatic pressing. Carter [38] et al. found that cracks in CM247LC alloy formed by SLM could be eliminated by hot isostatic pressing. Therefore, in order to eliminate the micro-cracks in as-EBM sample, selected PREP sample was given a HIP treatment. The SEM micrograph of the GH4099 after HIP indicates that HIP can eliminate the micro-cracks in as-EBMed GH4099. Figure 8a,b shows the microstructure evolution of γ' phase and the coarse irregular-shaped γ' phase after HIP appeared with the HIP temperature close to the initial melting point of GH4099, between 1423 K and 1453 K, which may affect the microstructure of the sample. Due to the fact that the hatching temperature of 1423 K is much higher than 1223 K in the process window, while the hatching time of a batch of samples is about 30 h, the achieved part retained its heat for a long time, which was equivalent to a high temperature solution treatment of the sample. After a long period of heat preservation, γ' phases were sufficiently precipitated and matured.

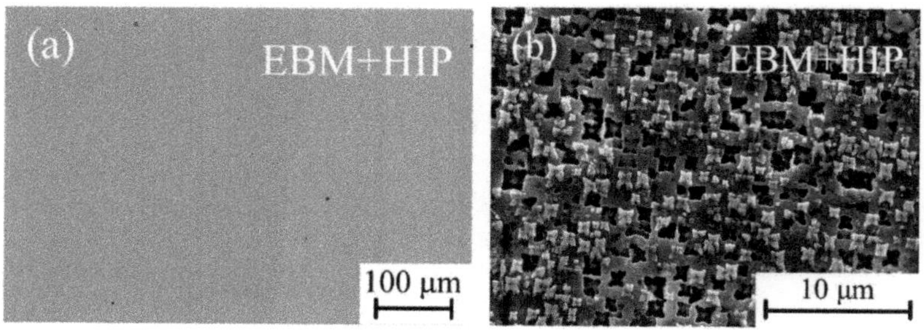

Figure 8. SEM micrographs of EBM+HIP PREP sample. Healed cracks after HIP (**a**) and the butterfly-like γ' precipitates (**b**).

Through various numerical simulations, Cha et al. [39] found that the growth rate in the <110> direction is enhanced due to a high chemical driving force, while the migration rate in the <100> direction is suppressed due to enriched solute and a corresponding low chemical driving force. The growing shape then transforms into a concave shape. The splitting occurs only in the concave growth condition and it is induced by interface instability, which is in good agreement with experimental observations. Yoo et al. [40] found that morphological unstable and irregular growth in Ni base superalloy occurs whenever there is enough supersaturation in the matrix so that the point effect of diffusion is operative with the change of morphology.

The resultant post-processed combined HIP with HT EBM GH4099 microstructure is illustrated in Figure 9. Heat treatment has a pronounced effect on the structure related to microfissuring through the effect on solute solubility, grain size, and/or the dissolution of phases [41]. Cracks were successfully closed and did not reopen after heat treatment (Figure 9a) which is different from our previous report on a nonweldable superalloy [13]. Figure 9b shows the features of γ' precipitates for heat-treated samples. The coarse irregular-shaped γ' was optimized into fine spherical particles and distributed more homogeneously in comparison with the HIP sample after heat treatment, which plays an effective role in adjusting the microstructure of the GH4099 alloy. The HIP+HT resulted in grain refinement, while γ' precipitates with more volume fraction. This was found after the HT when compared to that of the as-EBM material.

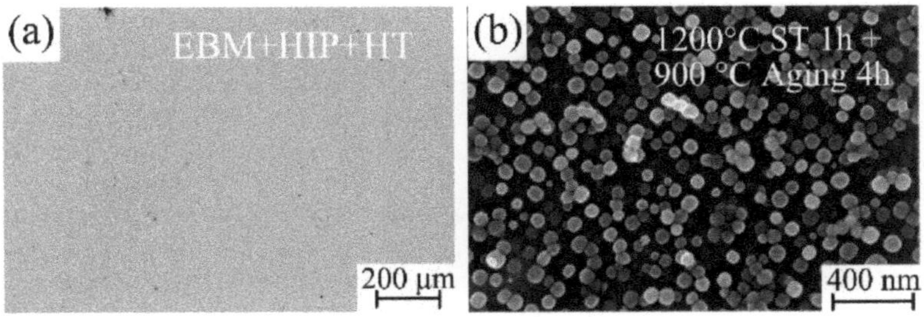

Figure 9. (**a**) The cracking incident after HIP+HT and (**b**) microstructural evolution in terms of γ' precipitates. The internal structure appears to be crack-free in EBM+HIP+HT condition.

Table 5 summarizes the grain width and γ' precipitates volume fraction of the GH4099 superalloy in different conditions. The results show that the difference of γ' volume fractions between GA and PREP powder-deposited GH4099 alloys with similar columnar crystal grain widths is small. When passed through HIP treatment of the PREPed sample, the grain width reduced to 70.74 ± 7.70 μm. The HIP temperature is 1423 K, which is close to the incipient melting temperature of the GH4099 alloy between 1423 K and 1453 K, thus affecting the microstructure of the sample. The 4 h heat preservation time during the HIP process is far less than the 30 h required for EBM formation of a batch of samples, and the grains had insufficient time to grow. When undergoing HT, the average grain width in the EBM+HIP condition of PREPed sample did not change significantly, but the number of γ' precipitates increased. These results show that PREPed sample undergoing HIP treatment then being exposed to 1473 K for 1 h + 1173 K for 4 h HT can not only completely eliminate the coarse irregular-shaped γ' precipitates, but also improve the mechanical properties of the alloy without affecting the grain structure of the superalloy.

Table 5. Measured average grain width and γ' volume fraction of as-EBM GAed and PREPed GH4099 samples and their changes due to HIP and HT. The building parameters of the GAed and PREPed sample are listed here.

Sample	I (mA)	V (mm/s)	L_{off} (mm)	Grain Width (μm)	γ' Volume Fraction (%)
GA-EBM	5	3000	0.1	195.16 ± 18.46	17.14 ± 0.92
PREP-EBM	10	1250	0.2	206.47 ± 28.74	16.10 ± 0.48
PREP-EBM+HIP	-	-	-	70.74 ± 7.70	-
PREP-EBM+HIP+HT	-	-	-	69.37 ± 6.32	19.30 ± 1.06

3.6. Mechanical Properties

The microhardness in the X-Z plane of GAed and PREPed samples under different conditions are compared with wrought GH4099 sample. The results show that the hardness value of the as-EBM GAed sample ~368 μm is equivalent to the wrought sample ~369 μm and is slightly lower than the hardness standard of the wrought GH4099 superalloy going through standard HT ~376 μm. The mechanical properties of GH4099 Ni-based superalloy in EBM condition have basically reached the forging standard. The microhardness of the PREPed GH4099 sample ~359 μm is slightly lower than that of wrought condition. As mentioned earlier, microcracks existed in the as-EBM PREPed specimens, and the hardness reflects the ability of the material to resist external force deformation. When measuring the microhardness at the same height of the prepared sample in a random manner, it is likely that the position of the measuring point is on or near the cracks, resulting in a decrease in the average value of the microhardness. At the same time, the standard deviation in microhardness measurement of the PREPed sample is also large, which may be due to inhomogeneous measuring points caused by the microcracks. The same case to GA sample is presumed as follows: GA powders were rapidly cooling and solidification was caused by argon gas during the preparation process, and there were more or less porosities inside. Therefore, the gas therein cannot escape in such a short time during a rapid solidification process, or the electron beam energy input is excessive, resulting in evaporation and ejection of the feedstock powders or molten materials, thereby generating pores in the deposited metals. Similarly, the uneven measuring points nearby porosities can lead to relatively big errors.

The microhardness in the X-Z plane of PREPed samples in EBM+HIP condition and EBM+HIP+HT condition was measured, compared with a wrought GH4099 sample after a standard HT. As shown in Figure 10, the microhardness of as-built PREPed sample decreases after the HIP treatment which is attributed to the closure of defects such as microcracks by the applied HIP treatment [42]. The γ' phase of the PREPed sample grew in an unstable manner, resulting in a large amount of a larger and irregular-shaped γ' phase. The enhancement in mechanical properties of a GH4099 Ni-based superalloy is mainly due to a precipitation strengthening effect with the main precipitates of γ'. It is beneficial to improve the mechanical properties as fine γ' precipitates dispersing uniformly in the γ matrix. The strengthening effect of a coarse irregular-shaped γ' phase relative to the γ matrix will be greatly weakened, leading to the decrease of microhardness. Subsequent HT of 1473 K solid solution for 1 h + 1173 K aging for 4 h caused the γ' phase to re-dissolve into the γ matrix in a fine dispersion during aging treatment.

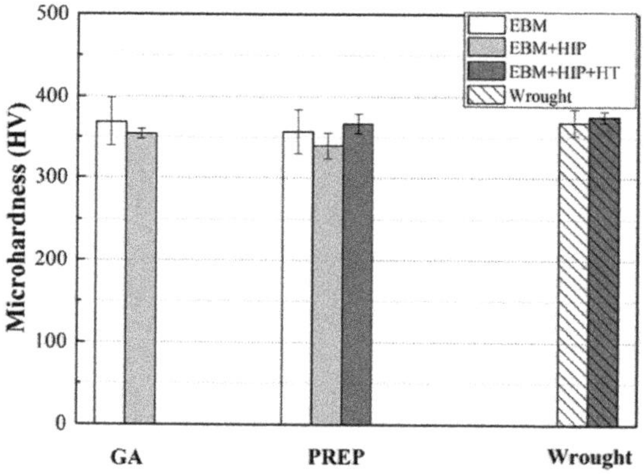

Figure 10. Measured average microhardness of GAed and PREPed GH4099 samples in EBM and EBM+HIP or EBM+HIP+HT condition compared with the wrought GH4099 in both as-wrought and wrought + HT conditions.

The elastic modulus and stacking fault energy between the two phases are different, which results in conformal strain and reinforcement effect on the alloy. Moreover, because of the larger specific area of the fine particles in the gamma phase, the reinforcement effect is significantly enhanced, which makes the PREP-EBM+HIP+HT specimens microscopic. Hardness has been improved. At the same time, cracks will not occur again after the solution aging treatment, so its microhardness is higher than that of the EBM state and EBM+HIP state. Meanwhile, the cracks did not re-generate inside the sample after HT, so its microhardness is notably enhanced compared with its in EBM and EBM+HIP conditions.

The as-EBM GAed sample was subjected to the same standard HT treatment. The strip tensile test specimen was cut parallel to the EBM building direction, its tensile strength at room temperature was measured, as well as in EBM+HIP+HT condition of PREPed samples which were cut parallel and perpendicular to the building direction, respectively. The above mentioned samples were compared with the tensile strength of a wrought GH4099 sample and by the standard HT at room temperature, as shown in the Figure 11.

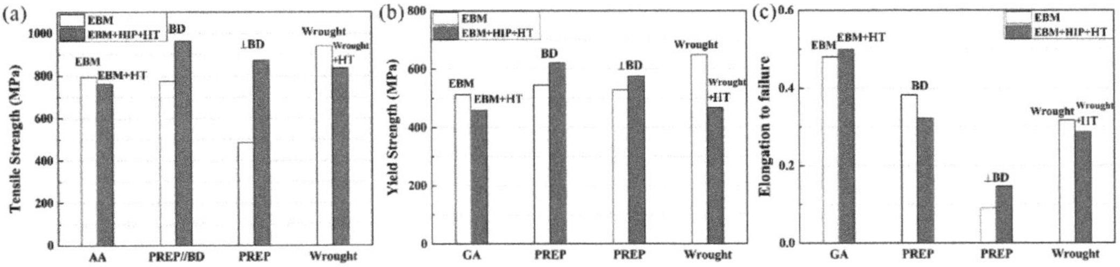

Figure 11. (**a**) Tensile strength, (**b**) yield strength, and (**c**) elongation to failure of GAed and PREPed GH4099 sample in different conditions cut parallel or/and perpendicular to building direction, as compared with that of a wrought sample along building direction.

Both of the tensile strengths of as-EBM GH4099 samples were lower than that of the wrought sample. Further, the tensile strength of the GAed sample is slightly higher than that of PREPed sample, which can be related to the small amount of pores inside GAed sample and the microcracks that existed in the PREPed sample. The tensile strength of post-processed PREPed sample along building direction was significantly more improved than the as-EBM sample due to the elimination of microcracks after HIP treatment and re-dissolving and re-precipitating dispersed γ' phase during HT, which provided a high mechanical strength of about ~926 μm. Tensile strength of the PREPed sample, which was cut perpendicular to the EBM building direction (i.e., perpendicular to the growth direction of the columnar crystal) at about ~486 μm was significantly lower than that parallel to the building direction about ~775 μm, which can be related to intergranular cracks distributed in as-EBM PREPed sample along grain boundary, resulting in low intergranular bonding force. Cracks rapidly expanded while stretching along building direction and then exhibited a low tensile strength. After HIP and HT, tensile strength increased to ~757 μm and was still slightly lower than the as-EBM PREPed tensile specimen parallel to the building direction indicating the anisotropy in strength. The tensile strength of the as-EBM GAed sample decreases after undergoing HT as is also the case for the wrought GH4099 sample.

4. Discussion

4.1. Crack Mechanism

GH4099 is considered difficult to weld due to its high cracking susceptibility related to high Al + Ti contents. The intergranular cracks were found particularly at columnar grain boundaries aligned along the building direction in as-EBM condition PREPed sample shown in Figure 7c, no transverse cracks and lack of fusion phenomenon were observed, indicating that the crack is an intergranular crack. Figure 12c,d show the secondary electron images of the crack surface from a PREPed sample which was cut into a 2 mm thickness sheet in the X-Z plane and pried apart running along the crack lines. Observation of the fracture surface shows the presence of discrete liquid films in quantities, which were wetting the dendrites when cracking occurred. It is clear that the high cooling rate for the EBM process leaves a magnitude of residual stresses, further supporting the notion of susceptibility of solidification cracking in presence of liquid films. On the other hand, thermal stresses during the last stage of solidification, pulling on the liquid films results in cracking [27]. The EBM melting process is quite a complicated course with re-melting, partial re-melting, cyclic annealing, etc. [27]. It is hard to identify the existing crack propagated as solidification cracking or liquation cracking. For solidification cracking, dendrite formation inhibits the flow of the remaining liquid in the interdendritic regions, which act as crack initiation points under the effect of the stress induced by solidification. Liquation cracking is generally reported to occur in a position away from the melt pool where the material is heated rapidly to a temperature which is lower than the overall liquidus of the material [43,44]. The fracture surface of crack-free GAed sample, on the other hand, exhibited dimple-like features as a response to a typical ductile failure shown in Figure 12b.

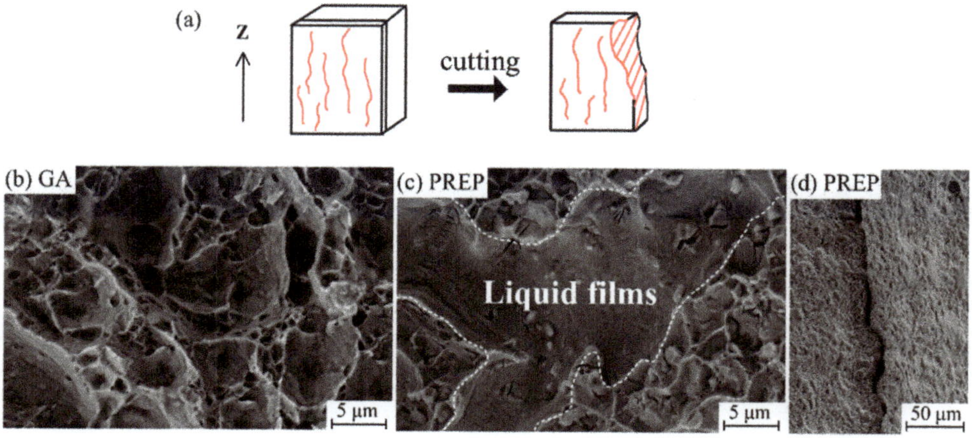

Figure 12. High magnification SE-SEM images of fracture surface from: (**b**) GAed sample, (**c**) a broken open PREPed sample, and (**d**) low magnification of (**c**). (**a**) Illustration of a PREPed sample that has been broken open.

Li et al. [45] reported that the crack initiated with the assistance of the transverse tensile strain/stress which tore up the liquid film formed by the low-melting point pre-existing phases in the primary heat affected zone, such as γ/γ' eutectics and coarse γ' precipitates. Thermal contraction and re-precipitation of secondary γ' leads to a high density of dislocations at the γ/γ' interface, further accelerating the separation of liquid film from grain boundary. Chauvet et al. [27] studied the fracture surface of a nonweldable Ni-based superalloy fabricated by a powder bed-based selective electron beam melting (S-EBM), and the dendritic morphology with a limited development of secondary arms indicated the presence of intergranular liquid films in the upper part. As the top layer had a limited shrinkage from former solidified layers, surface tensile effects were detrimental when tensile stress was applied to liquid film wetting dendrites. They also found that the cracks associated with the presence of liquid film propagated along high angle grain boundaries (misorientation > 15°), whereas low angle grain boundaries (misorientation < 15°) remained uncracked. Zhang et al. [46] established a mathematical model, which explained that different volume fractions of remaining liquid in CM247LC and IN792 Ni-based superalloys resulted in high strains and strain rates, thus affected crack formation and castability. The size of freezing range in the two-phase mushy zone contributes to the alloys property which decided hot tearing resistance, as per Refs. [46,47]. Moreover, hot tears occur when solid dendrite arms do not coalesce and liquid is still continuously present in between the dendrites above the coherency temperature, which is associated with deformation when tensile stresses are applied to the non-coherent dendrite network with insufficient liquid [47]. As for a nonweldable Ni-based superalloy, Kontis et al. [11] found that the amount of solutes segregation, namely of B, Mo, and Cr to HAGBs, led to the occurrence of liquid films where concentration in the liquid reached the critical composition to form borides along with the solidification stresses. They also identified hot cracking caused by grain boundary segregation, referred to as "segregation induced liquation", over the course of remelting and deposition of the subsequent layers in the additive manufacturing process.

4.2. Effect of Si Segregation on Cracking Behaviours

EPMA analysis represents the element distribution on fracture surface of PREPed sample in the X-Z plane. As shown in Figure 13, it can be observed that there are distinct Si and O enrichment at the crack edge, indicating that enrichment of trace elements such as Si contributes to the crack formation.

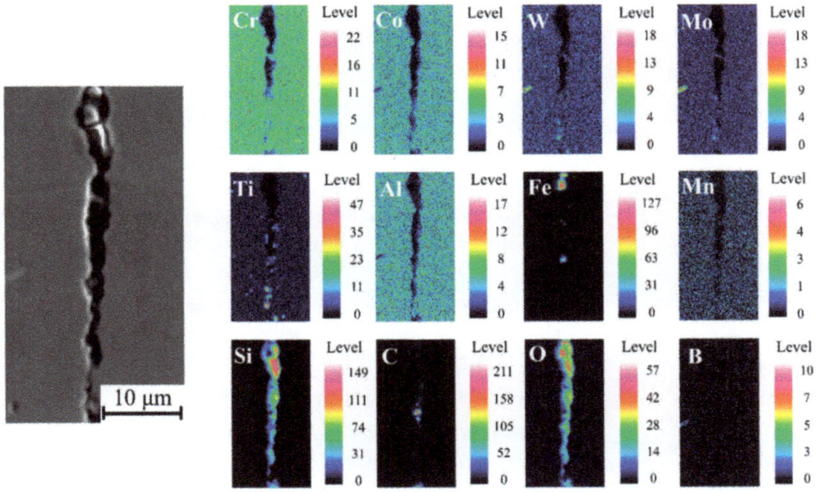

Figure 13. Series of elemental maps by EPMA showing the distribution near cracks of key elements in an as-EBM PREPed GH4099 superalloy. The colored scale bar at the right shows relative concentration, showing Si and O enrichment. Map conditions: 0.2 μm step size both in x and y, 20 kV.

Cloots et al. [48] revealed that cracks of IN738LC samples processed by selective laser melting propagate along building direction and distribute in a large grain boundary. Crack surfaces that seemed to be wetted with a liquid film were detected by atom probe tomography (APT), indicating the segregation of Zr at grain boundaries, which is a possible reason for hot cracking during the SLM process because it dramatically lowers the solidus temperature of IN738LC. Zhao et al. [49] investigated the fracture surface in René88DT superalloy prepared by laser solid forming (LSF) and formation of re-solidified products along grain boundaries in HAZ can be found. Energy dispersive spectroscopy (EDS) analysis results represented these re-solidified products within the cracks, containing high amount of Ti, Al, Cr, Co, and Ni, are the re-solidified (γ + γ') eutectics, and speculated that HAZ cracking results from liquation cracking. Engeli et al. [50] investigated crack density of IN738LC samples with different amounts of Si content fabricated by selective laser melting and found that the fraction of Si has a strong detrimental effect to the crack density even on a small amount <0.2 wt.%, as well as its SLM processability.

It is well known from previous works that minor elements including C, B, S, P, Mg, etc. strongly influence the weldability of Ni-based superalloys even on ppm level [34]. Silicon may be used as refining addition and deoxidizer to improve weldability and oxidation resistance during traditional melting process, but its detrimental effect presented in the final alloys must be considered according to Ref. [51].

ICP-AES results of chemical composition of the GA and PREP GH4099 powders in Table 1 show that the Si content of the PREP powder is higher by one order of magnitude than that of the GA powder. It can be seen that the GH4099 sample deposited with the PREP powder had a larger content of Si from vendors mixed in, thereby increasing the crack sensitivity during the EBM forming process, inducing the occurrence of intergranular crack.

5. Conclusions

GH4099 superalloy samples were successfully fabricated by EBM by using GA and PREP powders. Several observations and conclusions are drawn.

(1) Dense and crack-free samples were built at 1253 K for the GA powder. By contrast, cracks were observed in the PREP sample built at 1373 K. For both cases, fine spherical γ' phase precipitated uniformly with a volume fraction of ~20%.
(2) Local enrichment of minor element Si was responsible for the cracking behavior of the PREP sample.
(3) The cracks were successfully healed by HIP and heat treatment.
(4) Tensile properties of the GA and PREP samples in the build direction are comparable to a wrought superalloy. However, the healed cracks remain weak in the horizontal direction.

Author Contributions: S.W. conceived and designed the experiment, performed the experiments and wrote the paper; S.T. analyzed the data and discussed the results; H.P. designed the experiment and wrote the paper. All authors have read and agreed to the published version of the manuscript.

Funding: This research and the APC was funded by the National Science and Technology Major Project (J2019-VII-0016–0157) and National Key Research and Development Program of China (2021YFB3700501).

Data Availability Statement: Not applicable.

Conflicts of Interest: The authors declare no conflict of interest.

References

1. Altiparmak, S.C.; Xiao, B.W. A market assessment of additive manufacturing potential for the aerospace industry. *J. Manuf. Process.* **2021**, *68*, 728–738. [CrossRef]
2. Blakey-Milner, B.; Gradl, P.; Snedden, G.; Brooks, M.; Pitot, J.; Lopez, E.; Leary, M.; Berto, F.; du Plessis, A. Metal additive manufacturing in aerospace: A review. *Mater. Des.* **2021**, *209*, 110008. [CrossRef]
3. Yusuf, S.M.; Cutler, S.; Gao, N. Review: The Impact of Metal Additive Manufacturing on the Aerospace Industry. *Metals* **2019**, *9*, 1286. [CrossRef]
4. Linares, J.-M.; Chaves-Jacob, J.; Lopez, Q.; Sprauel, J.-M. Fatigue life optimization for 17-4Ph steel produced by selective laser melting. *Rapid Prototyp. J.* **2022**, *28*, 1182–1192. [CrossRef]
5. Yao, J.; Ding, R.C.; Li, K.L.; Du, B.R.; Zhao, L.; Yuan, Y.X. Study on the impact behavior of arch micro-strut (ARCH) lattice structure by selective laser melting (SLM). *Rapid Prototyp. J.* **2022**, *28*, 1541–1557. [CrossRef]
6. Panahizadeh, V.; Ghasemi, A.H.; Asl, Y.D.; Davoudi, M. Optimization of LB-PBF process parameters to achieve best relative density and surface roughness for Ti6Al4V samples: Using NSGA-II algorithm. *Rapid Prototyp. J.* **2022**. [CrossRef]
7. Gaur, B.; Soman, D.; Ghyar, R.; Bhallamudi, R. Ti6Al4V scaffolds fabricated by laser powder bed fusion with hybrid volumetric energy density. *Rapid Prototyp. J.* **2022**. [CrossRef]
8. Reed, R.C. *The Superalloys: Fundamentals and Applications*; Cambridge University Press: Cambridge, UK; New York, NY, USA, 2006.
9. Murray, S.P.; Pusch, K.M.; Polonsky, A.T.; Torbet, C.J.; Seward, G.G.E.; Zhou, N.; Forsik, S.A.J.; Nandwana, P.; Kirka, M.M.; Dehoff, R.R.; et al. A defect-resistant Co-Ni superalloy for 3D printing. *Nat. Commun.* **2020**, *11*, 1–11. [CrossRef]
10. Sotov, A.V.; Agapovichev, A.V.; Smelov, V.G.; Kokareva, V.V.; Dmitrieva, M.O.; Melnikov, A.A.; Golanov, S.P.; Anurov, Y.M. Investigation of the IN-738 superalloy microstructure and mechanical properties for the manufacturing of gas turbine engine nozzle guide vane by selective laser melting. *Int. J. Adv. Manuf. Technol.* **2020**, *107*, 2525–2535. [CrossRef]
11. Kontis, P.; Chauvet, E.; Peng, Z.; He, J.; da Silva, A.K.; Raabe, D.; Tassin, C.; Blandin, J.-J.; Abed, S.; Dendievel, R.; et al. Atomic-scale grain boundary engineering to overcome hot-cracking in additively-manufactured superalloys. *Acta Mater.* **2019**, *177*, 209–221. [CrossRef]
12. Yao, Y.; Xing, C.; Peng, H.; Guo, H.; Chen, B. Solidification microstructure and tensile deformation mechanisms of selective electron beam melted Ni3Al-based alloy at room and elevated temperatures. *Mater. Sci. Eng. A-Struct. Mater. Prop. Microstruct. Process.* **2021**, *802*, 140629. [CrossRef]
13. Peng, H.; Shi, Y.; Gong, S.; Guo, H.; Chen, B. Microstructure, mechanical properties and cracking behaviour in gamma'-precipitation strengthened nickel-base superalloy fabricated by electron beam melting. *Mater. Des.* **2018**, *159*, 155–169. [CrossRef]
14. Ramsperger, M.; Singer, R.F.; Korner, C. Microstructure of the Nickel-Base Superalloy CMSX-4 Fabricated by Selective Electron Beam Melting. *Metall. Mater. Trans. A-Phys. Metall. Mater. Sci.* **2016**, *47*, 1469–1480. [CrossRef]
15. Chauvet, E.; Tassin, C.; Blandin, J.-J.; Dendievel, R.; Martin, G. Producing Ni-base superalloys single crystal by selective electron beam melting. *Scr. Mater.* **2018**, *152*, 15–19. [CrossRef]
16. Fernandez-Zelaia, P.; Kirka, M.M.; Rossy, A.M.; Lee, Y.; Dryepondt, S.N. Nickel-based superalloy single crystals fabricated via electron beam melting. *Acta Mater.* **2021**, *216*, 117133. [CrossRef]

17. Kuo, Y.L.; Kamigaichi, A.; Kakehi, K. Characterization of Ni-Based Superalloy Built by Selective Laser Melting and Electron Beam Melting. *Metall. Mater. Trans. A-Phys. Metall. Mater. Sci.* **2018**, *49*, 3831–3837. [CrossRef]
18. Gaumann, M.; Henry, S.; Cleton, F.; Wagniere, J.D.; Kurz, W. Epitaxial laser metal forming: Analysis of microstructure formation. *Mater. Sci. Eng. A* **1999**, *271*, 232–241. [CrossRef]
19. Acharya, R.; Bansal, R.; Gambone, J.J.; Das, S. A Coupled Thermal, Fluid Flow, and Solidification Model for the Processing of Single-Crystal Alloy CMSX-4 Through Scanning Laser Epitaxy for Turbine Engine Hot-Section Component Repair (Part I). *Metall. Mater. Trans. B-Process Metall. Mater. Process. Sci.* **2014**, *45*, 2247–2261. [CrossRef]
20. Basak, A.; Acharya, R.; Das, S. Additive Manufacturing of Single-Crystal Superalloy CMSX-4 Through Scanning Laser Epitaxy: Computational Modeling, Experimental Process Development, and Process Parameter Optimization. *Metall. Mater. Trans. A-Phys. Metall. Mater. Sci.* **2016**, *47*, 3845–3859. [CrossRef]
21. Basak, A.; Das, S. Additive Manufacturing of Nickel-Base Superalloy Rene N5 through Scanning Laser Epitaxy (SLE)—Material Processing, Microstructures, and Microhardness Properties. *Adv. Eng. Mater.* **2017**, *19*, 1600690. [CrossRef]
22. Kirka, M.M.; Unocic, K.A.; Raghavan, N.; Medina, F.; Dehoff, R.R.; Babu, S.S. Microstructure Development in Electron Beam-Melted Inconel 718 and Associated Tensile Properties. *JOM* **2016**, *68*, 1012–1020. [CrossRef]
23. Korner, C.; Ramsperger, M.; Meid, C.; Burger, D.; Wollgramm, P.; Bartsch, M.; Eggeler, G. Microstructure and Mechanical Properties of CMSX-4 Single Crystals Prepared by Additive Manufacturing. *Metall. Mater. Trans. A-Phys. Metall. Mater. Sci.* **2018**, *49*, 3781–3792. [CrossRef]
24. DuPont, J.N.; Lippold, J.C.; Kiser, S.D. *Welding Metallurgy and Weldability of Nickel-Base Alloys*; John Wiley & Sons: Hoboken, NJ, USA, 2009.
25. Burger, D.; Parsa, A.B.; Ramsperger, M.; Korner, C.; Eggeler, G. Creep properties of single crystal Ni-base superalloys (SX): A comparison between conventionally cast and additive manufactured CMSX-4 materials. *Mater. Sci. Eng. A-Struct. Mater. Prop. Microstruct. Process.* **2019**, *762*, 138098. [CrossRef]
26. Wollgramm, P.; Burger, D.; Parsa, A.B.; Neuking, K.; Eggeler, G. The effect of stress, temperature and loading direction on the creep behaviour of Ni-base single crystal superalloy miniature tensile specimens. *Mater. High Temp.* **2016**, *33*, 346–360. [CrossRef]
27. Chauvet, E.; Kontis, P.; Jaegle, E.A.; Gault, B.; Raabe, D.; Tassin, C.; Blandin, J.-J.; Dendievel, R.; Vayre, B.; Abed, S.; et al. Hot cracking mechanism affecting a non-weldable Ni-based superalloy produced by selective electron Beam Melting. *Acta Mater.* **2018**, *142*, 82–94. [CrossRef]
28. Unocic, K.A.; Kirka, M.M.; Cakmak, E.; Greeley, D.; Okello, A.O.; Dryepondt, S. Evaluation of additive electron beam melting of haynes 282 alloy. *Mater. Sci. Eng. A-Struct. Mater. Prop. Microstruct. Process.* **2020**, *772*, 138607. [CrossRef]
29. Zhong, C.L.; Chen, J.; Linnenbrink, S.; Gasser, A.; Sui, S.; Poprawe, R. A comparative study of Inconel 718 formed by High Deposition Rate Laser Metal Deposition with GA powder and PREP powder. *Mater. Des.* **2016**, *107*, 386–392. [CrossRef]
30. Ding, P.P.; Mao, A.Q.; Zhang, X.; Jin, X.; Wang, B.; Liu, M.; Gu, X.L. Preparation, characterization and properties of multicomponent A1CoCrFeNi(2.1) powder by gas atomization method. *J. Alloys Compd.* **2017**, *721*, 609–614. [CrossRef]
31. Chen, Y.; Zhang, J.Y.; Wang, B.; Yao, C.G. Comparative study of IN600 superalloy produced by two powder metallurgy technologies: Argon Atomizing and Plasma Rotating Electrode Process. *Vacuum* **2018**, *156*, 302–309. [CrossRef]
32. Pleass, C.; Jothi, S. Influence of powder characteristics and additive manufacturing process parameters on the microstructure and mechanical behaviour of Inconel 625 fabricated by Selective Laser Melting. *Addit. Manuf.* **2018**, *24*, 419–431. [CrossRef]
33. Nguyen, Q.B.; Nai, M.L.S.; Zhu, Z.G.; Sun, C.N.; Wei, J.; Zhou, W. Characteristics of Inconel Powders for Powder-Bed Additive Manufacturing. *Engineering* **2017**, *3*, 695–700. [CrossRef]
34. Herzog, D.; Seyda, V.; Wycisk, E.; Emmelmann, C. Additive manufacturing of metals. *Acta Mater.* **2016**, *117*, 371–392. [CrossRef]
35. Helmer, H.E.; Korner, C.; Singer, R.F. Additive manufacturing of nickel-based superalloy Inconel 718 by selective electron beam melting: Processing window and microstructure. *J. Mater. Res.* **2014**, *29*, 1987–1996. [CrossRef]
36. Lopez-Galilea, I.; Ruttert, B.; He, J.Y.; Hammerschmidt, T.; Drautz, R.; Gault, B.; Theisen, W. Additive manufacturing of CMSX-4 Ni-base superalloy by selective laser melting: Influence of processing parameters and heat treatment. *Addit. Manuf.* **2019**, *30*, 100874. [CrossRef]
37. Kirka, M.M.; Medina, F.; Dehoff, R.; Okello, A. Mechanical behavior of post-processed Inconel 718 manufactured through the electron beam melting process. *Mater. Sci. Eng. A-Struct. Mater. Prop. Microstruct. Process.* **2017**, *680*, 338–346. [CrossRef]
38. Harrison, N.J.; Todd, I.; Mumtaz, K. Reduction of micro-cracking in nickel superalloys processed by Selective Laser Melting: A fundamental alloy design approach. *Acta Mater.* **2015**, *94*, 59–68. [CrossRef]
39. Cha, P.R.; Yeon, D.H.; Chung, S.H. Phase-field study for the splitting mechanism of coherent misfitting precipitates in anisotropic elastic media. *Scr. Mater.* **2005**, *52*, 1241–1245. [CrossRef]
40. Yoo, Y.S. Morphological instability of spherical gamma′ precipitates in a nickel base superalloy. *Scr. Mater.* **2005**, *53*, 81–85. [CrossRef]
41. Richards, N.L.; Chaturvedi, M.C. Effect of minor elements on weldability of nickel base superalloys. *Int. Mater. Rev.* **2000**, *45*, 109–129. [CrossRef]
42. Balachandramurthi, A.R.; Moverare, J.; Mahade, S.; Pederson, R. Additive Manufacturing of Alloy 718 via Electron Beam Melting: Effect of Post-Treatment on the Microstructure and the Mechanical Properties. *Materials* **2019**, *12*, 68. [CrossRef]
43. Dye, D.; Hunziker, O.; Reed, R.C. Numerical analysis of the weldability of superalloys. *Acta Mater.* **2001**, *49*, 683–697. [CrossRef]

44. Carter, L.N.; Attallah, M.M.; Reed, R.C. Laser Powder Bed Fabrication of Nickel-Base Superalloys: Influence of Parameters; Characterisation, Quantification and Mitigation of Cracking. In Proceedings of the 12th International Symposium on Superalloys, Seven Springs, PA, USA, 9–13 September 2012; pp. 577–586.
45. Li, Y.; Chen, K.; Tamura, N. Mechanism of heat affected zone cracking in Ni-based superalloy DZ125L fabricated by laser 3D printing technique. *Mater. Des.* **2018**, *150*, 171–181. [CrossRef]
46. Zhang, J.; Singer, R.F. Hot tearing of nickel-based superalloys during directional solidification. *Acta Mater.* **2002**, *50*, 1869–1879. [CrossRef]
47. Rappaz, M.; Drezet, J.; Gremaud, M. A New Hot-Tearing Criterion. *Metall. Mater. Trans. A-Phys. Metall. Mater. Sci.* **1999**, *30*, 449–455. [CrossRef]
48. Cloots, M.; Uggowitzer, P.J.; Wegener, K. Investigations on the microstructure and crack formation of IN738LC samples processed by selective laser melting using Gaussian and doughnut profiles. *Mater. Des.* **2016**, *89*, 770–784. [CrossRef]
49. Zhao, X.M.; Lin, X.; Chen, J.; Xue, L.; Huang, W.D. The effect of hot isostatic pressing on crack healing, microstructure, mechanical properties of Rene88DT superalloy prepared by laser solid forming. *Mater. Sci. Eng. A-Struct. Mater. Prop. Microstruct. Process.* **2009**, *504*, 129–134. [CrossRef]
50. Engeli, R.; Etter, T.; Geiger, F.; Stankowski, A.; Wegener, K. Effect of Si on the SLM processability of IN738LC. In *International Solid Freeform Fabrication Symposium*; The University of Texas: Austin, TX, USA, 2015; pp. 823–831.
51. Holt, R.T.; Wallace, W. Impurities and Trace Elements in Nickel-Base Superalloys. *Int. Met. Rev.* **1976**, *21*, 1–24. [CrossRef]

metals

Article

Effects of Process Parameters on the Relative Density and Properties of CuCrZr Alloy Produced by Selective Laser Melting

Jiaxin Li [1,†], Zezhou Kuai [1,†], Zhonghua Li [2,*], Bin Liu [1,*], Yanlei Chen [1], Shengyu Lu [1], Yunfei Nie [3] and Zhicheng Yang [2]

1 School of Materials Science and Engineering, North University of China, Taiyuan 030051, China; jiaxin_li_nuc@163.com (J.L.); kuaizezhou@163.com (Z.K.); chenyl778274787@163.com (Y.C.); lsy8819782@163.com (S.L.)
2 School of Mechanical Engineering, North University of China, Taiyuan 030051, China; yzc13653504211@163.com
3 State Key Laboratory of Mechanical Transmission, Chongqing University, Chongqing 400044, China; 20200701037@cqu.edu.cn
* Correspondence: lizhonghua6868@163.com (Z.L.); liubin3y@nuc.edu.cn (B.L.)
† These authors contributed equally to this work.

Abstract: CuCrZr alloy has the advantages of good electrical conductivity, thermal conductivity, high hardness, crack resistance and high softening temperature. It is extensively used in important fields such as rail transit, aerospace, thermonuclear fusion and electronic information. Due to its high melting point, reflectivity, thermal conductivity, etc., it is more difficult to manufacture by selective laser melting (SLM). In this work, the effect of SLM process parameters on the characteristics of CuCrZr samples, such as relative density, hardness and tensile properties, has been investigated using orthogonal experiment method. The experimental results show that laser power is the main factor affecting the properties of the alloy. The tensile strength of the alloy increases with an increase in laser power; it first increases and then decreases with an increase in scanning speed. The optimal combination of process parameters in this paper is as follows: laser power is 240 W, scanning speed is 750 mm/s and scanning spacing is 0.07 mm. The relative density, ultimate tensile strength and hardness of the alloy fabricated by best SLM process parameters was 98.79%, 347 MPa, 133.9 HV, respectively.

Keywords: selective laser melting; CuCrZr alloy; process parameters; relative density; tensile properties

Citation: Li, J.; Kuai, Z.; Li, Z.; Liu, B.; Chen, Y.; Lu, S.; Nie, Y.; Yang, Z. Effects of Process Parameters on the Relative Density and Properties of CuCrZr Alloy Produced by Selective Laser Melting. *Metals* **2022**, *12*, 701. https://doi.org/10.3390/met12050701

Academic Editor: Thomas Niendorf

Received: 15 March 2022
Accepted: 14 April 2022
Published: 19 April 2022

1. Introduction

Copper and its alloys have excellent electrical conductivity, thermal conductivity and corrosion resistance. They are widely used in electrical, metallurgy, machinery manufacturing, aerospace, defence industry and other fields, where they are made into lead frames for integrated circuits, overhead contact lines for electrified railways, radiators, aviation thrust chambers, etc. [1–3]. Copper alloys are mostly prepared by traditional casting processes. However, it is difficult to fabricate parts with complex structures and internal runners, so that greatly limit the application of copper alloys. Three-dimensional printed copper alloy parts have good mechanical properties, excellent microstructure and surface quality that make them easier to polish. They are widely used in jewellery and cultural education and can also be used to manufacture micro radiators and aerospace engine combustion chamber components.

Selective laser melting (SLM) is an additive manufacturing process. According to the Computer Aided Design (CAD) 3D model, the laser melts the powder layer by layer along the path planned by the path planning software. Due to its high thermal conductivity and electrical conductivity, the laser powder bed melting of copper and copper alloys has become an increasingly popular topic. However, copper alloys have excellent thermal conductivity and higher reflectivity. This makes it difficult for lasers to continuously melt the powder, which leads to difficulties in melting copper alloys and controlling the quality of powder metallurgy. Concurrently, higher laser power will cause certain damage to the laser device. Huang et al. [4] prepared pure copper parts using SLM technology and studied the influence of the process on the microstructure evolution and mechanical properties of pure copper parts. Lodes et al. [5] used electron beam melting (EBM) technology to manufacture pure copper parts with a relative density of up to 99.5%. Tian et al. [6] fabricated Cu-13.5Al-4Ni-0.5Ti copper-based shape memory alloy using SLM technology. By optimising the parameters, an almost completely dense copper-based sample was obtained, and its phase and microstructure were characterised.

With the development of the electronics industry, the need for high-strength and high-thermal-conductivity copper alloys is increasing, and copper–chromium alloys are increasingly favoured as typical precipitation-strengthened copper alloys. CuCrZr has excellent mechanical properties and electrical conductivity, alongside thermal conductivity, so it has broad development and application prospects [7–9]. The study of CuCrZr alloys prepared using SLM technology has become a research hotspot. Ma [10] established a statistical model of the influence of processing parameters on relative density using the responding surface method and analysis of variance, obtained the optimal SLM process parameters of CuCrZr alloy and prepared nearly fully dense samples. The microstructure and mechanical properties of the sample were studied. Compared with deformed copper alloys, SLMed CuCrZr has a considerable tensile strength (~321 MPa), but its elongation at break (25%) is much higher than that of forged samples (15%). Sun et al. [11] used an ANSYS simulation to determine the approximate process range of the CuCrZr alloy produced by SLM. Then, the effects of laser power, scanning speed and hatching distance on the mechanical properties of CuCrZr alloy in the range of SLM process parameters were investigated by experiments. When the laser power was 460 W, the scanning speed was 700 mm/s and the hatching distance was 0.06 mm, the optimal performance of the alloy was obtained. The tensile strength was 153.5 MPa, the hardness was 119 HV, and the roughness was 31.384 μm. Guan et al. [12] studied the effect of heat treatment on CuCrZr alloy manufactured by SLM. As the ageing temperature increased, the strength of the alloy gradually decreased, and the electrical conductivity gradually increased. After ageing at 500 °C, the strength and electrical conductivity of the alloy reached 490 MPa and 73% IACS, respectively. Wallis et al. [13] studied CuCrZr samples fabricated using SLM and discussed the influence of heat treatment on the microstructure and properties of the samples. Compared with conventional heat treatment, direct age-hardening improved the mechanical properties, and clusters of Cr and Cu_xZr_y particles were detected in the sample. Hu et al. [14] studied the changes in the microstructure and mechanical properties of Cu-Cr-Zr alloys prepared by selective laser melting (SLM) with different scanning parameters. By optimising the scanning parameters, a CuCrZr specimen with a relatively high relative density (99.5 $\pm$ 0.3%) was obtained, with a strength of (280 $\pm$ 6 MPa) and a plasticity of (23.4 $\pm$ 0.4%). However, the mechanical properties of CuCrZr alloy prepared by SLM in these studies are lower than forging. Therefore, it is necessary to improve the mechanical properties and processing efficiency of SLMed CuCrZr alloy by adjusting laser power, scanning speed, hatching distance, scanning strategy, etc.

In the study, the CuCrZr alloy samples were fabricated through SLM to determine which of the process parameters were optimised by orthogonal experiments. The effects of scanning power, scanning speed and hatching distance on the properties of SLM CuCrZr alloy samples were analysed.

2. Materials and Methods

2.1. Experimental Materials and Equipment

The CuCrZr powder produced by gas atomisation was used in the SLM manufacturing process, and the chemical composition is shown in Table 1. The size distribution of CuCrZr powder was measured by Mastersizer (S3500, Microtrac, Osaka, Japan), its particle size distribution ranged from 15.25 μm (D10) to 46.59 μm (D90), with an average particle size of 29.98 μm, as shown in Figure 1. A SU5000 (Hitachi, Tokyo, Japan) SEM was used to characterise the microstructure of the CuCrZr powder and observe the fracture morphology of tensile parts. Figure 2 shows the microstructure of the CuCrZr powder. Most of the powder particles are spherical or nearly spherical, with a smooth surface and good fluidity that is beneficial to the flatness of the powder layer to ensure the formation of SLM. A small part of the powder particles presents irregular shapes, and the small powder particles are adsorbed around the large powder particles. EP-M150 (E-PLUS 3D, Beijing, China) equipment was used to process CuCrZr alloy samples. It is equipped with a 500 W Yb: YAG laser with a wavelength of 1064 nm and a spot size of 70 μm with a Gaussian distribution. To reduce the temperature gradient, the substrate preheating temperature was 100 °C, the powder layer thickness was 30 μm, and a scanning strategy of 67° rotation per layer was adopted. During the printing process, the oxygen content of the moulding chamber was less than 0.02% and protected by argon gas.

Table 1. Chemical compositions of CuCrZr alloy powders.

Element	Cr	Zr	Cu
Content (wt. %)	0.75	0.077	Bal.

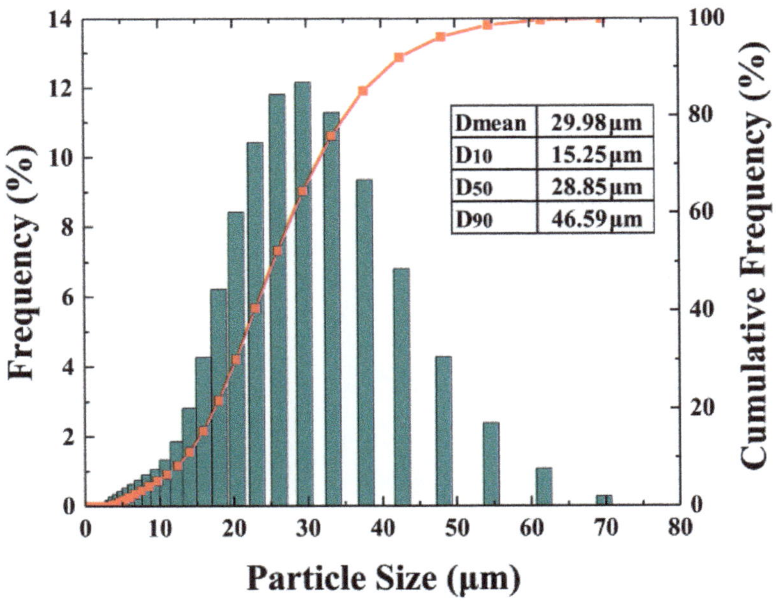

Figure 1. Particle size distribution of CuCrZr powder.

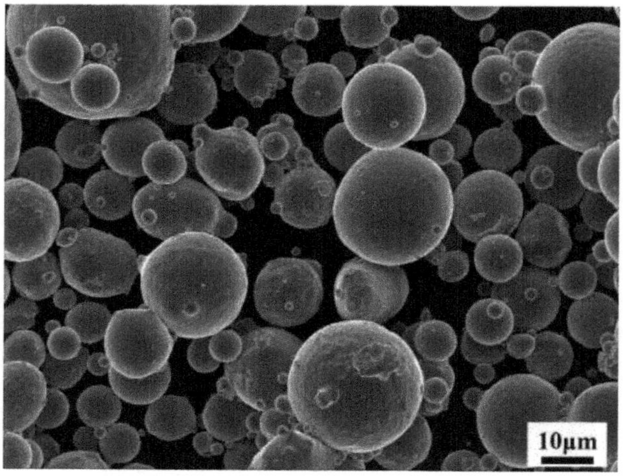

Figure 2. SEM image of CuCrZr powder.

2.2. Experimental Design

Since there are many factors affecting the forming quality of SLM, it will be a heavy workload to carry out research in full factor mode. Therefore, this experiment adopts a 3-factor 4-level orthogonal test method [15]. The three factors selected in this paper are laser power, scanning speed and hatching distance, mainly because these three factors have great influence on forming quality when the same SLM equipment prepares parts. In order to avoid large workload and more reliable test results, Taguchi L16 was used to generate general results which have four levels for each factor. The process parameters used are shown in Table 2. In this experiment, an orthogonal test was used to prepare the CuCrZr sample with a size of 10 mm × 10 mm × 10 mm, as shown in Figure 3.

Table 2. Process parameters of alloys fabricated by SLM.

Group	Test Number	Laser Power P (W)	Scanning Speed V (mm/s)	Hatching Distance D (μm)
A	No. 01	160	650	50
	No. 02	160	750	60
	No. 03	160	850	70
	No. 04	160	950	80
B	No. 05	200	650	60
	No. 06	200	750	50
	No. 07	200	850	80
	No. 08	200	950	70
C	No. 09	240	650	70
	No. 10	240	750	80
	No. 11	240	850	60
	No. 12	240	950	50
D	No. 13	280	650	80
	No. 14	280	750	70
	No. 15	280	850	60
	No. 16	280	950	50

Figure 3. CuCrZr samples fabricated by SLM.

2.3. *Experimental Feature*

Completed samples were removed from the 304 stainless steel base plate via wire electrical discharge machining (Wire-EDM). The separated samples were polished with sandpaper until the surface was bright and without scratches. A Zeiss optical microscope (Axio Scope A1, Carl Zeiss AG, Oberkochen, Germany) was used to shoot the pores of the cross-section. Image J—an image-processing software (V 1.8.0, National Institutes of Health, Bethesda, MD, USA)—was used for relative density measurement. At least 20 pictures were measured for each sample, and the average value was taken. The Vickers hardness tester (JMHVS-1000AT, Shanghai Material testing Machine Factory, Shanghai, China) was used to test the Vickers hardness of the samples: we pressed it into the surface of the samples with a test force of 200 g and kept it there for 10 s. Then we measured 10 indentations for each sample and took the average value as the final result. The electronic universal testing machine (AG—X PLUS, Shimadzu Corporation, Kyoto, Japan) was used to complete the test of the tensile sample; the tensile speed was 1 mm/min, and the stress–strain data were collected before the sample was broken. The same kinds of tensile specimens were tested three times, and the results were averaged.

3. Results and Discuss

3.1. Relative Density

In order to ensure the accuracy of relative density calculation, no less than 20 optical images were collected for each sample. After removing the highest and lowest values, the average value was taken as the final result. Figure 4 shows the relative density measurement results and statistical errors of volume fraction of the pores driven from the image analysis of 16 samples manufactured by SLM. It is obvious from Figure 4 that when laser power is 240 W, scanning speed and hatching distance have little influence on relative density. The relative density of sample No. 15 is the highest, reaching 98.98%. This is mainly because at this process parameter, there is enough energy to completely melt the powder and enough time for the gas inside the powder bed to escape, etc.

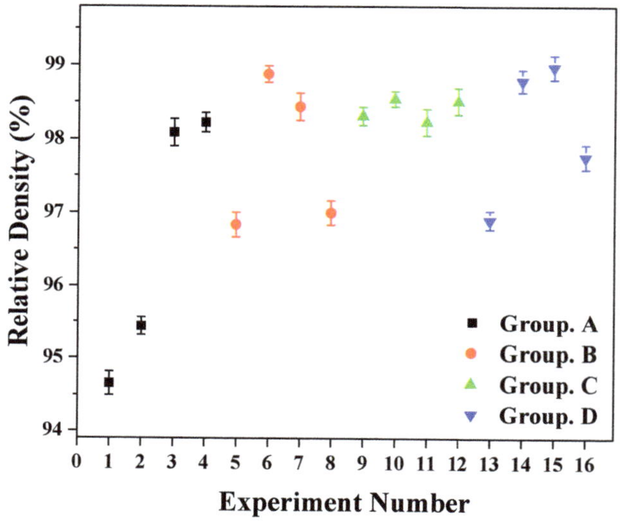

Figure 4. Relative density of samples fabricated by SLM with different process parameters.

Figure 5 is the main effect diagram of alloy relative density under various process parameters obtained through orthogonal analysis. Figure 5a shows that when the laser power is increased from 160 W to 280 W, the compactness of the samples first shows a trend of increasing, then decreasing. When the laser power is too low, the energy absorbed by the powder becomes too low to be completely melted, the viscosity of the melt becomes higher and the fluidity becomes poor, resulting in poor lap quality between the melt channels and poor continuity of the melt channels. In addition, there are more irregular pores in the sample where the powder is not completely fused, resulting in a low relative density of the sample. When the laser power is 200 W, the sample relative density is significantly improved, as shown in Figure 6a,b. When the laser power rises to 280 W, the powder absorbs so much energy that the melt begins to evaporate. Fast expansion of vapor leads to explosions in the boiling melt [16]. The splashed particles adhere to the surface of the sample and cause a poor weld lap. Excessive laser power can cause spheroidisation and even burn. In addition, a large amount of molten metal evaporates, resulting in many circular pores on the surface of the sample, thereby reducing the relative density of the sample. When the laser power is high, the solidification time of the molten pool is prolonged, the hydrogen in the molten pool reaches the local solubility limit and nucleation generates bubbles. The bubbles diffuse from the bottom of the molten pool and the heat-affected zone to the top and eventually form many hydrogen holes, so the hole defects are obvious [17,18], as shown in Figure 6c,d.

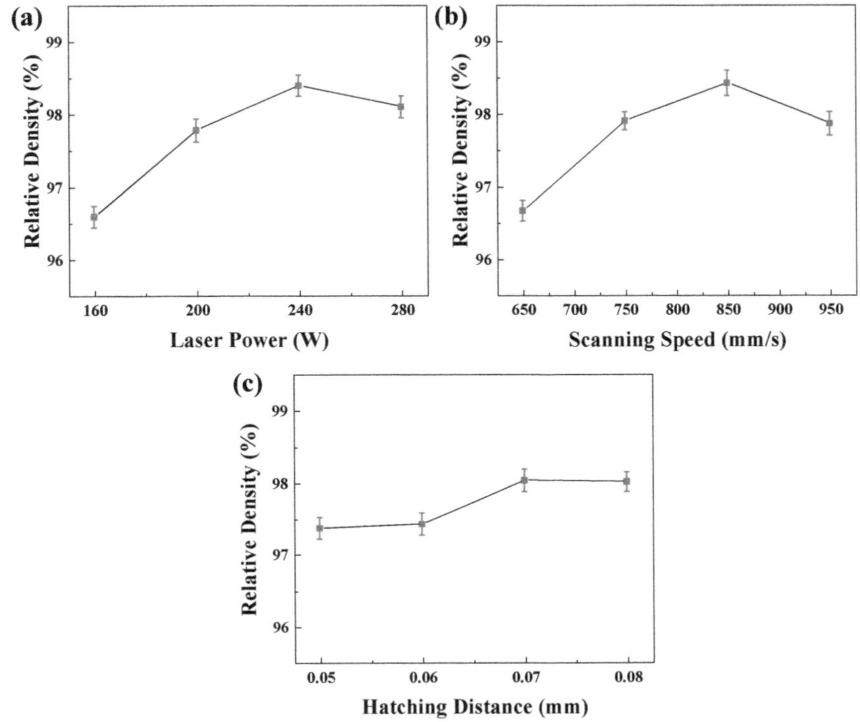

Figure 5. The influence of process parameters on the relative density of samples: (**a**) Laser Power, (**b**) Scanning Speed, (**c**) Hatching Distance.

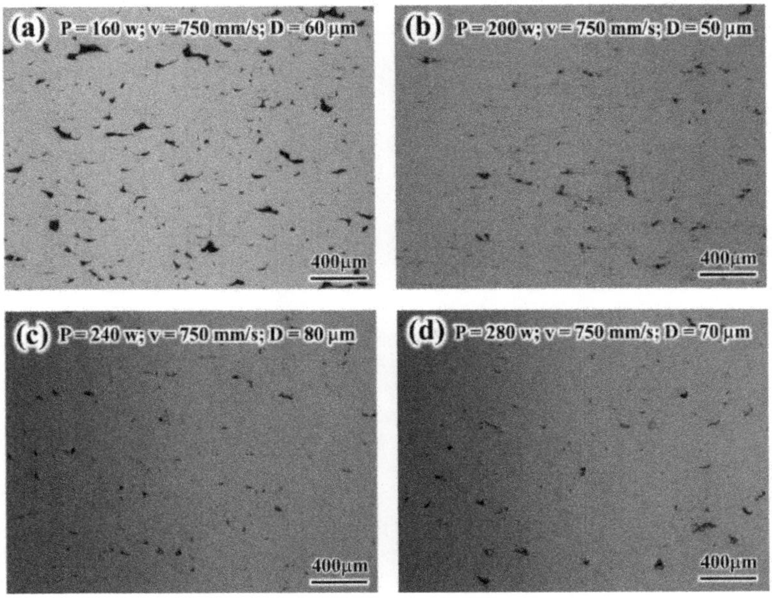

Figure 6. Optical micrograph of sample: (**a**) No. 2, (**b**) No. 6, (**c**) No. 10, (**d**) No. 14.

As the scanning speed increases, the relative density of the sample slowly increases and then decreases, as shown in Figure 5b. When the scanning speed is too low, it is easy to make the local liquid phase too much, resulting in nodules and decreased relative density [17,18]. Concurrently, when V is too small, due to the strong Marangoni effect, defects such as spheroidisation and pores are easily formed in the molten pool, so the overall relative density value is not high. When the scanning speed is 950 mm/s, the high scanning speed makes the cooling rate of the molten pool faster, the width and depth of the molten pool are reduced and the melting channel is prone to discontinuity, which leads to a decrease in relative density.

Figure 5c shows that as the hatching distance increases, the relative density of the alloy first increases and then decreases. The smaller the hatching distance, the greater the internal stress and the greater the temperature gradient, which is likely to cause spheroidisation and even lead to warpage deformation of the samples. Since the melting channel is elliptical during the forming process, the two melting channels are distributed in a wave shape. When the hatching distance is too large, the overlap area between adjacent melting channels is reduced, resulting in a large amount of unmelted powder in the trough area and forming many pores of very large size. A reasonable hatching distance can make the powder in the overlapping area between adjacent melting channels fully melt, and the molten metal can be completely spread, reducing the generation of pores. The body energy density formula can be used to express the influence on density. The formula is as follows:

$$E = \frac{P}{vdh} \tag{1}$$

where E is the energy density of the laser body, P is the laser power, v is the scanning speed, d is the hatching distance and h is the thickness of the powder layer.

Table 3 shows the results of the analysis of the influence of process parameters on relative density. The p-value of the laser power is the smallest, indicating that the laser power is the main factor affecting the relative density, followed by the scanning speed, and the hatching distance has the least influence.

Table 3. Results from the regression analysis of the data.

Source	DF	Adj SS	Adj MS	F-Value	p-Value
Laser power	3	9.964	3.321	2.94	0.052
Scanning speed	1	5.278	5.278	4.67	0.109
Hatching distance	1	3.387	3.387	3.00	0.305
Error	12	13.565	1.130	-	-
Total	15	23.529	-	-	-

3.2. *Effect of Process Parameters on the Microhardness of SLM Alloy*

The main effects plot showing the variation of the hardness of SLM alloy with laser power, scanning speed and hatching distance is plotted in Figure 7. When the laser power is 280 W, the average hardness of the sample reaches 135.3 HV0.2. The laser energy density increased with the increased of laser power, which resulted in strong metallurgical bonding between powder particles [19]. When the laser power is 160 W, the average hardness of the sample becomes the lowest because when the sample gets less energy, the powder cannot be fully melted, and the fluidity of the metal solution becomes poor, resulting in the sample having more pores and lower relative density. When the scanning speed is 750 mm/s, the hardness of the sample becomes higher. CuCrZr is a precipitated hardened copper alloy [20,21]. Under the action of pressure, precipitated Cr atoms hinder the sliding between grains, thus improving the hardness of the alloy [11]. However, with the increase in scanning speed, the hardness of samples decreased, which was mainly caused by Cr atoms in solid solution.

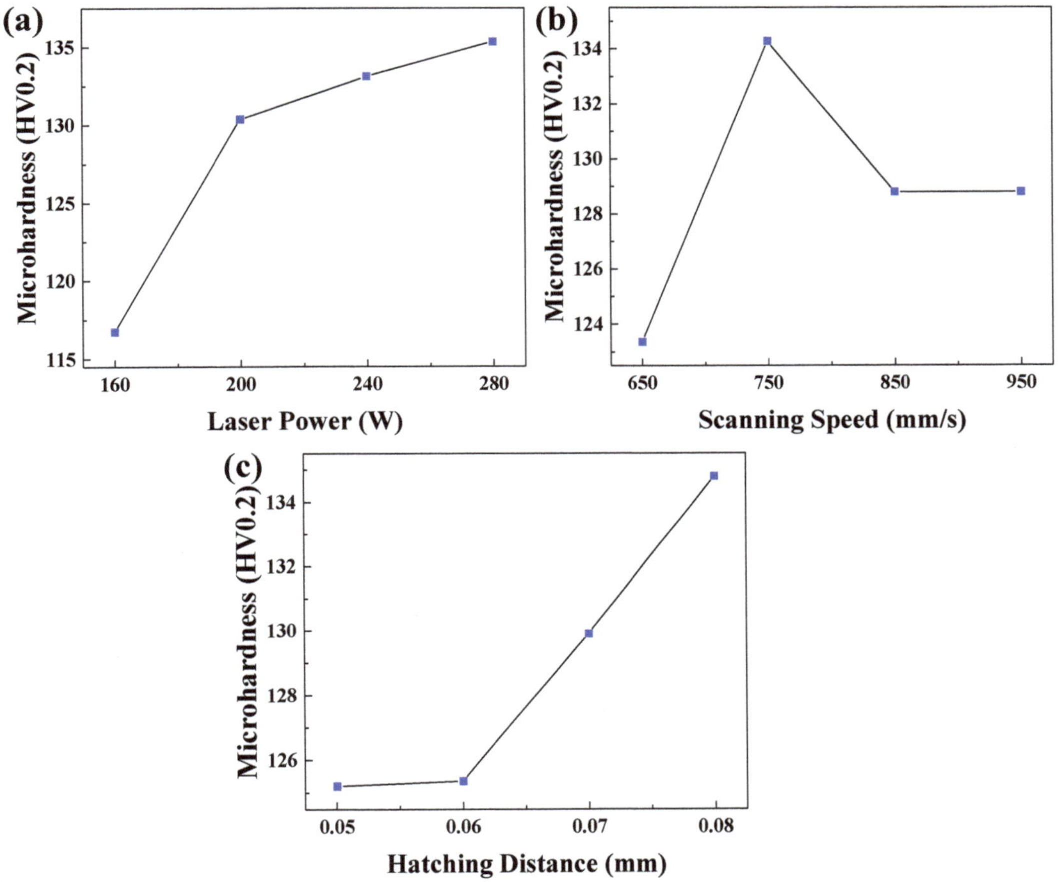

Figure 7. The influence of process parameters on the hardness of samples: (**a**) Laser Power, (**b**) Scanning Speed, (**c**) Hatching Distance.

Table 4 shows the average hardness value of the upper surface of the sample. From the table, the hardness of the Group A samples is lower. As the laser power increases, the hardness value of the alloy increases, and rate of increase decreases.

Table 4. The average hardness value of the upper surface of the sample.

Experiment No.	Microhardness (HV0.2)	Experiment No.	Microhardness (HV0.2)
1	97.4 ± 1.56	9	139.9 ± 2.39
2	125.7 ± 0.85	10	138.4 ± 1.91
3	120.9 ± 1.06	11	125.5 ± 0.14
4	122.9 ± 0.35	12	128.7 ± 1.48
5	117.3 ± 0.38	13	138.9 ± 2.12
6	139.1 ± 0.57	14	133.8 ± 1.77
7	138.9 ± 2.97	15	129.8 ± 1.27
8	124.8 ± 0.64	16	138.8 ± 0.99

3.3. Strength of the SLMed Alloy

Figure 8 shows the main effect diagram of the influence of the process parameters on the tensile properties of the samples. When the laser power is 280 W, the tensile and yield strengths of the sample become the largest at 339 MPa and 153 MPa, respectively. With a laser power of 160 W, the tensile and yield strengths of the sample are the smallest. This is due to the increase in void defects at higher scanning speeds, which intensifies the initiation and propagation of cracks. Figure 8b shows that as the scanning speed increases, the alloy's tensile and yield strengths show a slow downwards trend. This is because the increase in scanning speed leads to instability of liquid relative flow, imbalance of the interface and crystal failure. Figure 8c shows that as the hatching distance increases, the alloy's tensile and yield strengths decrease, but the decrease is smaller. Thus, the hatching distance has little effect on the alloy's tensile and yield strengths. In summary, there are differences in the properties of CuCrZr alloys prepared with different process parameters.

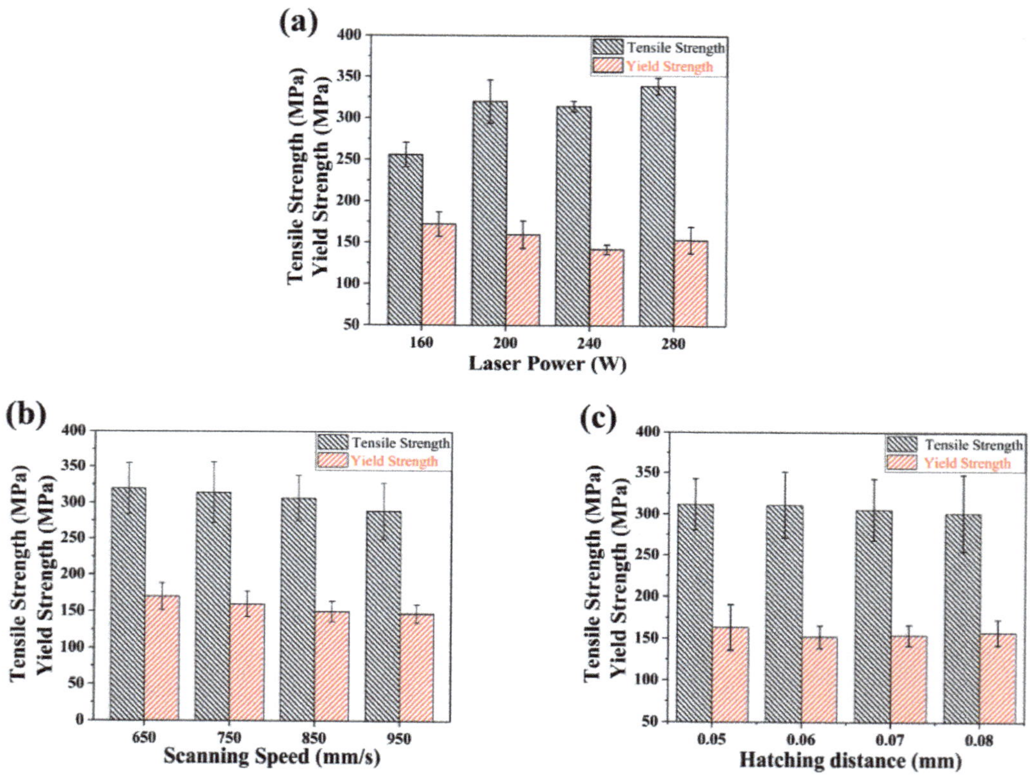

Figure 8. The influence of process parameters on the tensile properties of samples: (**a**) Laser Power, (**b**) Scanning Speed, (**c**) Hatching Distance.

Figure 9 shows the stress–strain curves of CuCrZr alloys prepared by SLM under different process parameters. It can be seen from the figure that the CuCrZr alloy has no obvious yield stage. When the load reaches the maximum, the sample breaks immediately. Sample No. 4 has the lowest energy density and the lowest tensile strength, with a tensile strength of 235 MPa and an elongation of 6.91%. Sample No. 14 has the highest tensile strength, with a tensile strength of 347 MPa and an elongation of 17.26%. It can be seen from Figure 9a that when the laser power is 160 W, the tensile strength of the SLM formed sample is low, and the elongation is small. As the laser power increases, the tensile strength and elongation of the pattern generally increase. Sample No. 16 has the highest energy density, and compared with other samples in group D, the tensile strength and elongation are significantly reduced. The ultimate tensile strength and elongation generally show a positive growth trend, that is, the greater the ultimate tensile strength, the greater the elongation.

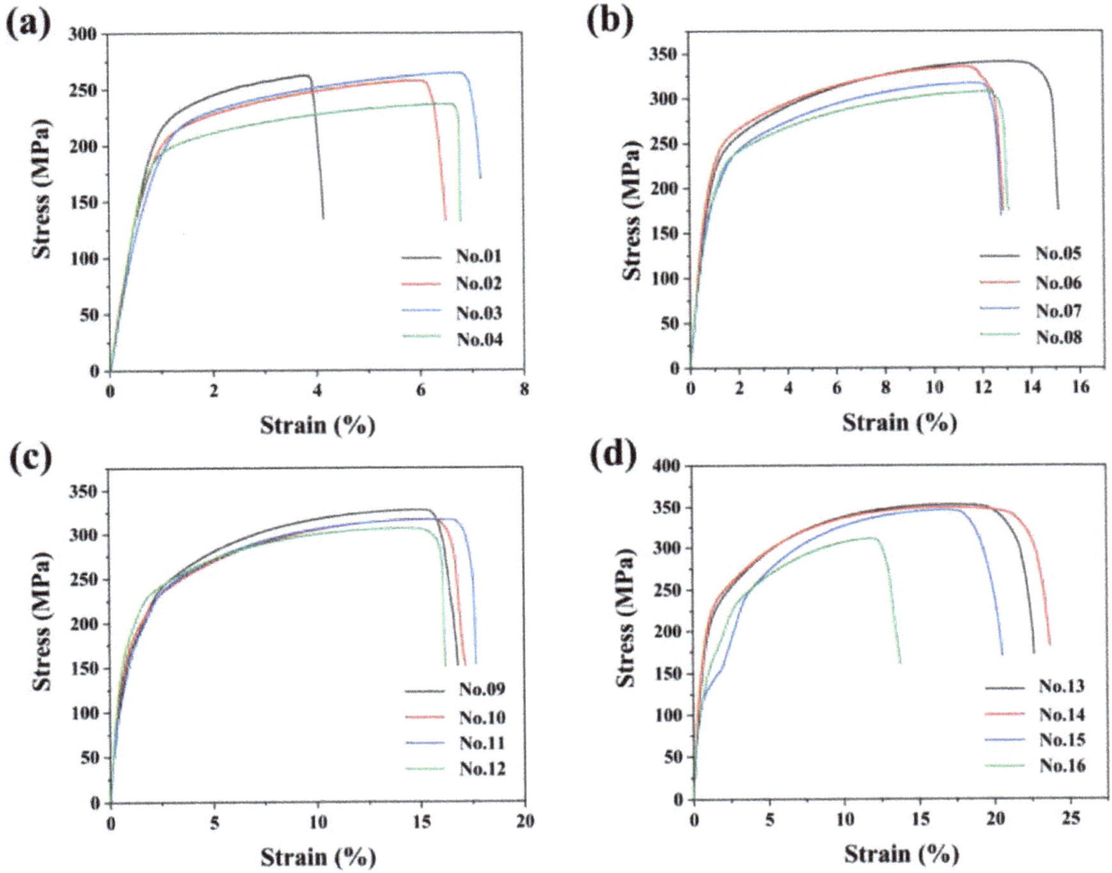

Figure 9. Stress–strain curve of CuCrZr alloy fabricated by SLM under different process parameters. (**a**) Group A, (**b**) Group B, (**c**) Group C, (**d**) Group D.

Figure 10 shows a typical SEM image of the fracture of the CuCrZr alloy tensile sample. Figure 10a–c show the fracture morphologies of sample No. 4 at low-magnification and high-magnification, respectively. Figure 10a shows the poor metallurgical bonding of sample No. 4 with many large irregular sizes defects on the surface of the tensile fracture. There are many unmelted spherical powder particles in the interior and peripheral areas of the pores, as shown in Figure 10b. Figure 10d–f are the low-magnification and high-magnification fracture morphologies of Sample No. 14, respectively. The tensile fracture shows an obvious necking phenomenon, and many micron-scale dimples appear, showing the characteristics of ductile fracture. The structure shows that an increase in laser power can significantly increase the tensile strength of the sample and significantly increase ductility.

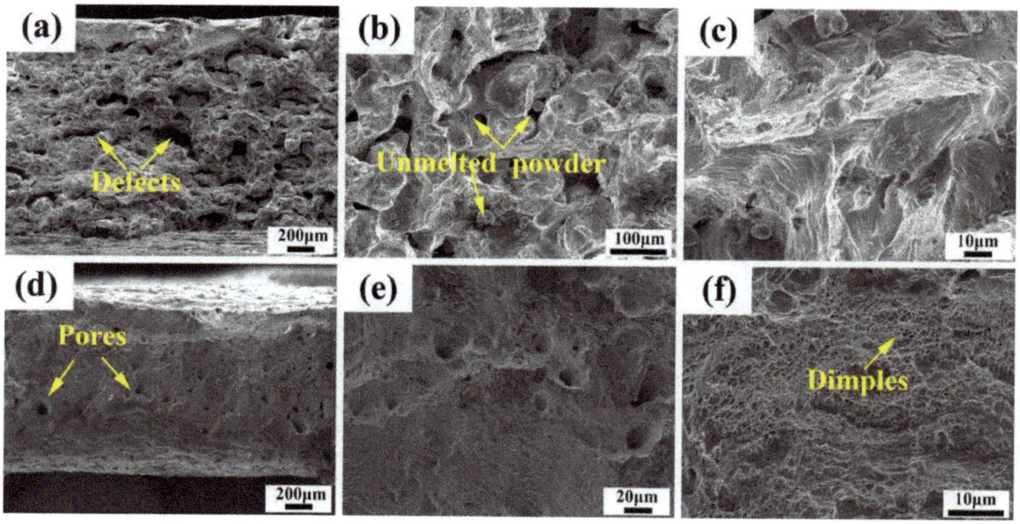

Figure 10. SEM images of the fracture morphologies of the tensile parts from (**a**–**c**) No.4 and (**d**–**f**) No.14.

4. Conclusions

In this paper, the effects of process parameters (laser power, scanning speed and hatching distance) on the relative density, hardness and tensile strength of SLM alloys were studied through orthogonal experiments, and CuCrZr alloys with good comprehensive properties were prepared. The main conclusions are as follows:

When the energy density is low, there are many irregular pores on the surface of the tensile fracture, and a large number of unmelted spherical powder particles exist in and around the pores. With the increase in energy density, the overall trend of sample relative density is to increase first, remain stable, and then slowly decrease. The laser power is the main factor that affects relative density, followed by scanning speed, and hatching distance has the least influence.

The hardness of the alloy keeps rising, and the rate of increase decreases with the increase in laser power. With the increase in scanning speed, the hardness of alloy increases, and the increase rate decreases. The tensile strength and elongation of the alloy increase with the laser power. When the laser power is 280 W, the scanning speed is 750 mm/s, and the hatching distance is 0.07 mm; the alloy has the greatest ultimate tensile strength and the highest elongation. The tensile fracture shows an obvious necking phenomenon, and many micron-scale dimples appear, showing the characteristics of ductile fracture.

The optimal combination of process parameters in this paper is as follows: laser energy is 240 W, scanning speed is 750 mm/s and hatching distance is 0.07 mm. Under these parameters, the ultimate tensile strength and elongation of CuCrZr alloy are the highest, reaching 347 MPa and 17.26%, respectively. However, its relative density is still lower than that of Ti6Al4V and 316L prepared by SLM technology. Therefore, future work will focus on improving the relative density and mechanical properties of CuCrZr alloy through post-treatment.

Author Contributions: Writing—original draft, investigation, formal analysis, J.L.; writing—review and editing and supervision, Z.K.; methodology, supervision and resource, Z.L.; methodology and resource, B.L.; conceptualization and methodology, Y.C.; methodology, S.L.; investigation, writing—review and editing, Y.N.; writing—review and methodology, Z.Y. All authors have read and agreed to the published version of the manuscript.

Funding: This research was funded by the NSFC (Grant No. 51905497, No. 52075502), the Major Science and Technology Projects of Shanxi Province, China (Grant No. 20181102012) and the Support Program for Young Academic Leaders of North University of China (Grant No. QX201902). The APC was funded by the NSFC (Grant No. 52075502).

Institutional Review Board Statement: Not applicable.

Informed Consent Statement: Not applicable.

Data Availability Statement: Additional supporting data can be provided upon request.

Conflicts of Interest: The authors declare no conflict of interest.

References

1. Wang, W.; Zhu, J.; Qin, N.; Zhang, Y.; Li, S.; Xiao, Z.; Lei, Q.; Li, Z. Effects of minor rare earths on the microstructure and properties of Cu-Cr-Zr alloy. *J. Alloy. Compd.* **2020**, *847*, 155762. [CrossRef]
2. Junfeng, W.; Jinshui, C.; Chengjun, G.; Jianbo, Z.; Xiangpeng, X.; Bin, Y. Effect of heat treatment on low cycle fatigue properties of Cu–Cr–Zr alloy. *Mater. Charact.* **2019**, *158*, 109940. [CrossRef]
3. Huang, A.H.; Wang, Y.F.; Wang, M.S.; Song, L.Y.; Li, Y.S.; Gao, L.; Huang, C.X.; Zhu, Y.T. Optimizing the strength, ductility and electrical conductivity of a Cu-Cr-Zr alloy by rotary swaging and aging treatment. *Mater. Sci. Eng. A* **2019**, *746*, 211–216. [CrossRef]
4. Huang, J.; Yan, X.; Chang, C.; Xie, Y.; Ma, W.; Huang, R.; Zhao, R.; Li, S.; Liu, M.; Liao, H. Pure copper components fabricated by cold spray (CS) and selective laser melting (SLM) technology. *Surf. Coatings Technol.* **2020**, *395*, 125936. [CrossRef]
5. Lodes, M.A.; Guschlbauer, R.; Körner, C. Process development for the manufacturing of 99.94% pure copper via selective electron beam melting. *Mater. Lett.* **2015**, *143*, 298–301. [CrossRef]
6. Tian, J.; Zhu, W.; Wei, Q.; Wen, S.; Li, S.; Song, B.; Shi, Y. Process optimization, microstructures and mechanical properties of a Cu-based shape memory alloy fabricated by selective laser melting. *J. Alloy. Compd.* **2019**, *785*, 754–764. [CrossRef]
7. Kalinin, G.M.; Ivanov, A.D.; Obushev, A.N.; Rodchenkov, B.S.; Rodin, M.E.; Strebkov, Y.S. Ageing effect on the properties of CuCrZr alloy used for the ITER HHF components. *J. Nucl. Mater.* **2007**, *367–370*, 920–924. [CrossRef]
8. Wang, W.; Li, R.; Zou, C.; Chen, Z.; Wen, W.; Wang, T.; Yin, G. Effect of direct current pulses on mechanical and electrical properties of aged Cu–Cr–Zr alloys. *Mater. Des.* **2016**, *92*, 135–142. [CrossRef]
9. Zhang, B.; Zhang, Z.-G.; Li, W. Mechanical properties, electrical conductivity and microstructure of CuCrZr alloys treated with thermal stretch process. *Trans. Nonferrous Met. Soc. China* **2015**, *25*, 2285–2292. [CrossRef]
10. Ma, Z.; Zhang, K.; Ren, Z.; Zhang, D.Z.; Tao, G.; Xu, H. Selective laser melting of Cu–Cr–Zr copper alloy: Parameter optimization, microstructure and mechanical properties. *J. Alloys Compd.* **2020**, *828*, 154350. [CrossRef]
11. Sun, F.; Liu, P.; Chen, X.; Zhou, H.; Guan, P.; Zhu, B. Mechanical Properties of High-Strength Cu–Cr–Zr Alloy Fabricated by Selective Laser Melting. *Materials* **2020**, *13*, 5028. [CrossRef] [PubMed]
12. Guan, P.; Chen, X.; Liu, P.; Sun, F.; Zhu, C.; Zhou, H.; Fu, S.; Wu, Z.; Zhu, Y. Effect of selective laser melting process parameters and aging heat treatment on properties of CuCrZr alloy. *Mater. Res. Exp.* **2019**, *6*, 1165c1. [CrossRef]
13. Wallis, C.; Buchmayr, B. Effect of heat treatments on microstructure and properties of CuCrZr produced by laser-powder bed fusion. *Mater. Sci. Eng. A* **2018**, *744*, 215–223. [CrossRef]
14. Hu, Z.; Du, Z.; Yang, Z.; Yu, L.; Ma, Z. Preparation of Cu–Cr–Zr alloy by selective laser melting: Role of scanning parameters on densification, microstructure and mechanical properties. *Mater. Sci. Eng. A* **2022**, *836*, 142740. [CrossRef]
15. Khorasani, A.M.; Gibson, I.; Awan, U.S.; Ghaderi, A. The effect of SLM process parameters on density, hardness, tensile strength and surface quality of Ti-6Al-4V. *Addit. Manuf.* **2019**, *25*, 176–186. [CrossRef]
16. Wen, P.; Jauer, L.; Voshage, M.; Chen, Y.; Poprawe, R.; Schleifenbaum, J.H. Densification behavior of pure Zn metal parts produced by selective laser melting for manufacturing biodegradable implants. *J. Mater. Processing Technol.* **2018**, *258*, 128–137. [CrossRef]
17. Weingarten, C.; Buchbinder, D.; Pirch, N.; Meiners, W.; Wissenbach, K.; Poprawe, R. Formation and reduction of hydrogen porosity during selective laser melting of AlSi10Mg. *J. Mater. Processing Technol.* **2015**, *221*, 112–120. [CrossRef]
18. Pal, S.; Lojen, G.; Kokol, V.; Drstvensek, I. Evolution of metallurgical properties of Ti-6Al-4V alloy fabricated in different energy densities in the Selective Laser Melting technique. *J. Manuf. Processes* **2018**, *35*, 538–546. [CrossRef]
19. Chen, J.; Yang, Y.; Song, C.; Wang, D.; Wu, S.; Zhang, M. Influence mechanism of process parameters on the interfacial characterization of selective laser melting 316L/CuSn10. *Mater. Sci. Eng. A* **2020**, *792*, 139316. [CrossRef]
20. Singh, R.; Singh, S.; Kanigalpula, P.; Saini, J. Electron beam welding of precipitation hardened CuCrZr alloy: Modeling and experimentation. *Trans. Nonferrous Met. Soc. China* **2020**, *30*, 2156–2169. [CrossRef]
21. Feng, H.; Jiang, H.; Yan, D.; Rong, L. Microstructure and Mechanical Properties of a CuCrZr Welding Joint After Continuous Extrusion. *J. Mater. Sci. Technol.* **2015**, *31*, 210–216. [CrossRef]

metals

MDPI

Article

Effect of Solution Temperature on the Microstructure and Properties of 17-4PH High-Strength Steel Samples Formed by Selective Laser Melting

Zhanyong Zhao [1,*], Hailin Wang [1], Pengcheng Huo [1], Peikang Bai [2,*], Wenbo Du [3], Xiaofeng Li [1], Jianmin Li [4] and Wei Zhang [4]

1 School of Materials Science and Engineering, North University of China, Taiyuan 030051, China; whj671424473@126.com (H.W.); nuchpc@126.com (P.H.); lxf@nuc.edu.cn (X.L.)
2 College of Materials Science and Engineering, Taiyuan University of Science and Technology, Taiyuan 030024, China
3 National Key Laboratory for Remanufacturing, Academy of Army Armored Forces, Beijing 100072, China; dwbneu@163.com
4 Technical Center, Taiyuan Iron and Steel (Group) Co., Ltd., Taiyuan 030003, China; lijm@tisco.com.cn (J.L.); zhangwei@tisco.com.cn (W.Z.)
* Correspondence: zhaozy@nuc.edu.cn (Z.Z.); baipeikang@nuc.edu.cn (P.B.)

Abstract: Samples of 17-4PH high-strength steel were processed by selective laser melting (SLM) and solution-processed. The effects of the solution temperature on the microstructure and mechanical properties of the samples were studied. The 17-4PH high-strength steel is primarily composed of martensite, with a small number of austenite phases, and contains many dislocations. After the solution treatment, the grain size gradually increased, yielding typical martensite. The samples were subjected to an aging treatment after the solution treatment. Precipitates formed in the samples, conducive to improving their strength and hardness. The Vickers hardening and wear properties of the 17-4PH high-strength steel samples first increased and then decreased with increasing solution temperature. After the solution treatment at 1040 °C for 2 h and aging at 480 °C for 4 h, the Vickers hardening of the 17-4PH high-strength steel increased to 392 HV0.5, and the friction coefficient was approximately 0.6. These values were, respectively, 7% and 5% higher than those for the untreated samples.

Keywords: 17-4PH; SLM; solution temperature; microstructure and properties

Citation: Zhao, Z.; Wang, H.; Huo, P.; Bai, P.; Du, W.; Li, X.; Li, J.; Zhang, W. Effect of Solution Temperature on the Microstructure and Properties of 17-4PH High-Strength Steel Samples Formed by Selective Laser Melting. *Metals* **2022**, *12*, 425. https://doi.org/10.3390/met12030425

Academic Editor: Thomas Niendorf

Received: 21 January 2022
Accepted: 25 February 2022
Published: 28 February 2022

1. Introduction

The 17-4PH high-strength steel is a low-carbon martensitic stainless steel containing nickel and copper that can be hardened by precipitation [1]. This stainless steel material has high strength and strong corrosion resistance [2,3]. It is widely used in medical devices, the chemical industry, automobile manufacturing, the military industry, the aerospace industry, and the nuclear industry. Owing to the melting point of this material, it can only be used for a long time at temperatures below 30 °C [4,5]. The intended application conditions of this steel evolve and become more stringent; consequently, the performance requirements become more demanding. The martensitic structure and the high hardness caused by precipitation hardening increase the processing difficulty of this material [6,7]. The traditional high-temperature processing technology cannot meet the requirements of the processing of parts, such as no casting process defects, and some complex parts [8]. This negatively affects subsequent deformation processing and heat treatment; as a result, complex welded parts can easily accumulate stress. Therefore, selective laser melting (SLM) forming technology was proposed in an attempt to address the forming problem [9]. SLM uses a metal powder that melts and cools rapidly under the heat of a laser beam, which overcomes the problems caused by traditional methods of manufacturing metal parts with complex shapes [10] The 17-4PH high-strength steel relies on the precipitation of the second phase to strengthen the material [11,12]. Martensitic transformation and aging treatment in order to form the precipitation hardening phase are the main strengthening methods [13] Along with SLM, solution treatment and aging treatment of 17-4PH high-strength steels also importantly affect SLM-based sample formation, with improved performance under application and service conditions [14].

Yadollahi et al. [15] studied the effects of heat treatment and crack orientation on the fatigue crack growth (FCG) behavior of 17-4PH stainless steel samples fabricated using a laser powder bed fusion (LPBF) additive manufacturing (AM) system. Their experimental results showed that following heat treatment, a large amount of delamination was observed in the crack propagation region of the specimens with transverse cracks. When the crack tip encountered a poorly bonded layer, a sudden deflection of the propagating crack also occurred in the specimens with longitudinal cracks, which were formed due to insufficient fusion or due to high residual stress. Qi et al. [14] studied the solution and aging treatments of 17-4PH stainless steel samples in a vacuum and under hot isostatic pressing (HIPA, HIPB, HIPC) conditions; the samples were prepared using SLM. The relative density, tensile strength, and ductility improved following the solution and aging treatments performed under the HIP conditions. Sun et al. [16] studied the effects of conventional solution and precipitation hardening (H-900) heat treatments on the microstructure evolution of 17-4PH AM and wrought components. These microstructural studies demonstrated that microstructures and hardnesses similar to those of wrought samples could be achieved in 17-4PH AM components by post-build heat treatments. Cheruvathur et al. [17] compared 17-4PH AM with forging 17-4 PH in conventional production. The effects of post-build thermal processing on the material microstructure were compared to those for conventionally produced wrought 17-4 PH with the intention of creating a more uniform, fully martensitic microstructure. Chung and Tzeng [18] studied the effect of aging treatment at 480 °C, 520 °C, and 620 °C on the microstructure and mechanical properties of metal injection molding (MIM) 17-4PH. A large number of fine spherical ε-Cu precipitates on the lath martensite matrix used to strengthen the MIM 17-4PH effectively prevented the movement of dislocations and yielded the highest hardness after aging at 480 °C.

However, there are only a few systematic studies on the effect of solution temperature on the microstructure and properties of 17-4PH. In this study, 17-4PH high-strength steel samples were prepared by SLM and then heat-treated to improve their mechanical properties. The effects of the solution temperature on the microstructure, hardness, friction, and wear and tensile properties of the SLM-formed 17-4PH samples were studied.

2. Experimental Procedures

The main chemical components (wt. %) of the 17-4PH high-strength steel powder are listed in Table 1. During the experiment, 17-4PH high-strength steel powder was screened, impurities in the powder were filtered out, and the filtered powder was put into a drying box for drying treatment. After drying, there was no moisture in the metal powder. In the experiment, EP-M150 selective laser melting forming equipment (Beijing Eplus 3D Tech. Co., Ltd, Beijing, China) was used. The instrument was equipped with a 200 W/500 W water-cooled fiber laser with a laser wavelength of 1060–1090 nm, and the protective atmosphere was argon. A chessboard-laser-scan strategy was adopted, and a laser focus diameter at the powder bed of 70 μm was used. Moreover, the SLM process parameters of the laser power (P), the scanning speed (v), the hatch spacing (h), and the layer thickness (t) were used in the experiment, and their values are listed in Table 2.

Table 1. Chemical components (wt. %) of the 17-4PH high-strength steel powder.

Cr	Ni	Cu	Si	Mn	Nb	C	Fe
14.10	3.94	5.64	0.47	0.37	0.21	0.07	balance

Table 2. SLM processing parameters used to produce the specimens.

Laser Power (W)	Scanning Speed ($mm{\cdot}s^{-1}$)	Layer Thickness (μm)	Hatch Spacing (μm)	Energy Density (J/mm^3)
185	620	30	110	90.4

Moreover, the starting temperature Ms of the transformation from the austenite structure to the martensite structure is above room temperature. After heat treatment, all the structures of the sample become martensite. However, when the solution temperature of the heat treatment is too high, the martensite structure is uneven, the Ms point decreases, the residual austenite increases, and the ferrite content also increases. All these make the heat treatment effect worse. Therefore, the heat treatment system (solution-treated at 940 °C, 990 °C, 1040 °C, 1090 °C, and 1140 °C for 2 h) was selected based on comprehensive consideration of these factors, and then the samples were quenched in water. After this process, the samples were aged at 480 °C for 4 h. The heat treatment system is presented in Table 3.

Table 3. The heat treatment system for the SLM-formed 17-4PH high-strength steel.

Number	Heat Treatment System
#1	Solution 940 °C × 2 h (water cooling) + Aging 480 °C × 4 h (air cooling)
#2	Solution 990 °C × 2 h (water cooling) + Aging 480 °C × 4 h (air cooling)
#3	Solution 1040 °C × 2 h (water cooling) + Aging 480 °C × 4 h (air cooling)
#4	Solution 1090 °C × 2 h (water cooling) + Aging 480 °C × 4 h (air cooling)
#5	Solution 1140 °C × 2 h (water cooling) + Aging 480 °C × 4 h (air cooling)

The phase identification of the specimens' surfaces (specimens' dimensions: 10 mm × 10 mm × 10 mm) was performed using X-ray diffractometry (XRD) (D/max-Rb; $K\alpha1$, 0.15408 nm; acceleration voltage, 40 kV; current, 100 mA; diffraction angle range, 20–110°; scanning speed, 3°/min; Beijing Zhongke light analysis Chemical Technology Research Institute, Beijing, China). The specimens were polished and etched with a solution of 4 g $CuSO_4$ + 20 mL HCl + 20 mL H_2O. The microstructure and distribution of elements in the samples were examined through the use of a scanning electron microscope (SEM, JSM-7900F, JEOL Co., Ltd., Tokyo, Japan) equipped with an energy-dispersive spectroscope (EDS, JEOL Co., Ltd., Tokyo, Japan). The samples were processed to a thickness of 30 μm and were further thinned using a precision ion-polishing system. Transmission electron microscopy (TEM, JEM-2100F, JEOL Co., Ltd., Tokyo, Japan) observations were performed using a field emission gun operating at an accelerating voltage of 200 kV. The hardening behavior was measured using a Vickers hardness tester (HVS-1000, Shanghai Yanrun Light-Machine Technology Co., Ltd., Shanghai, China) under a load of 0.5 N and a holding time of 15 s. The specimen surfaces were kept flat and smooth and were assessed at 10 equidistant points. The mechanical properties of the samples were measured using an Instron 3382 mechanical properties testing system at room temperature (the strain rate was 1 mm/min, and an overall outline of the dimensions of the tensile specimen is shown in Figure 1). An HRS-2M high-speed reciprocating friction and wear tester (Lanzhou Zhongke Kaihua Technology Development Co., Ltd., Lanzhou, China) was used for non-lubricated friction and wear tests at room temperature. The load was 30 N, the friction speed was 300 t/s, the friction length was 5 mm, and the friction pair was a Si_3N_4 ceramic ball with a diameter of 0.4 mm.

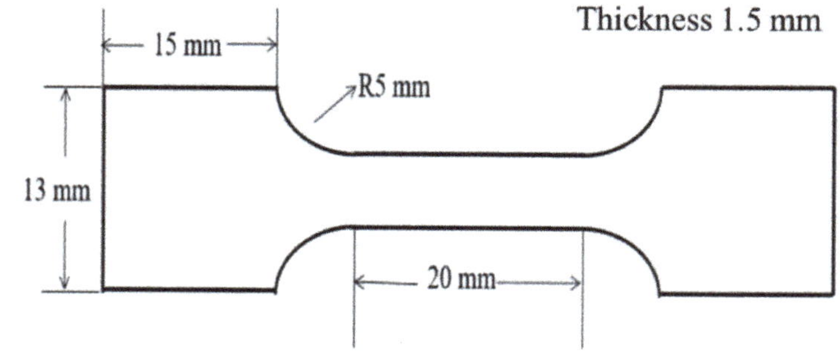

Figure 1. An overall outline of the dimensions of the tensile specimen.

3. Results and Discussion

3.1. Effect of the Solution Temperature on the Microstructure

The TEM morphological images of the 17-4PH high-strength steel samples formed by SLM are shown in Figure 2. As the TEM sample becomes thinner, perforations tend to form in the fine grain at the boundary of the molten pool. Examining the TEM images, the samples' structure mainly included lath martensite with the body-centered cubic (BCC) crystal structure and austenite grain with the face-centered cubic (FCC) crystal structure. As shown in Figure 2a, the width of the lath martensite was 200–300 nm. As shown in Figure 2b, the trace width of residual austenite along the boundary of the martensite lath was 200–300 nm. The crystal structure of the martensite was BCC, and a high density of dislocations was observed as shown in Figure 2c. Moreover, as shown in Figure 2d, many spherical inclusions (with particle sizes in the 50–170 nm range) were also present in the TEM images. According to the EDS composition analysis of the precipitates, the contents of the O and Si elements were high, and these spherical inclusions were determined to be Si-rich oxides.

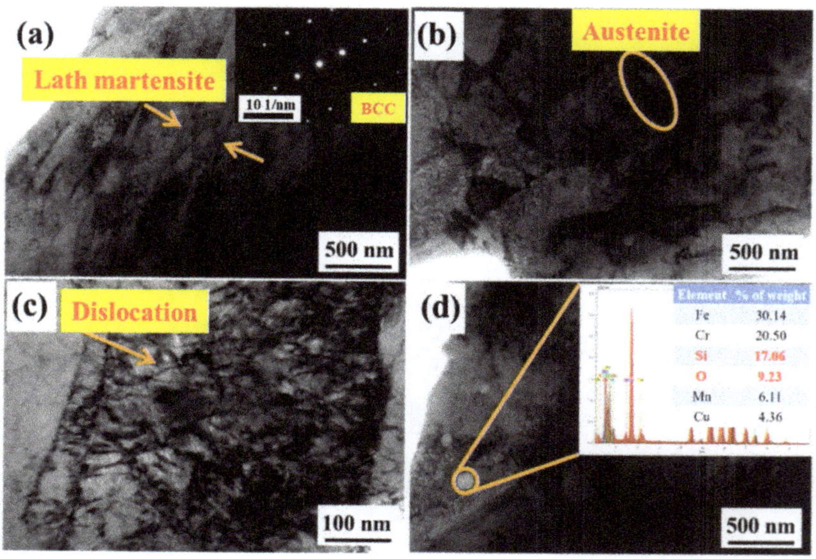

Figure 2. TEM images of the morphology of the 17-4PH high-strength steel formed by SLM: (**a**) martensite; (**b**) austenite; (**c**) dislocation; (**d**) precipitates.

Barroux et al. [19] mention that the formation of these inclusions in the forming process is considered a manufacturing defect, despite the use of protective gas. Si-rich oxide inclusions were formed during the SLM-based production. Such inclusions, considered as manufacturing defects, were reported by several authors for 316 L stainless steel [20,21].

Figures 3 and 4 show the SEM morphology and XRD patterns of the 17-4PH high-strength steel samples formed by SLM-based processing and subsequent heat treatment. Because the transformation point temperatures of 17-4PH high-strength steel are 670 °C for Ac1, 740 °C for Ac3, 140 °C for Ms, and 32 °C for Mf, the best performance requirements can be obtained by adjusting different heat treatment processes. Figure 3a shows a sample that was formed using SLM, and the structure is composed of lath martensite. Corresponding to the XRD pattern shown in Figure 3b, the structure also features the γ-austenite peak. After the heat treatment, the samples gradually formed a typical martensite microstructure as the temperature of the solution treatment increased. As shown in Figure 3c, the grain size increased as the solution temperature increased up to 940 °C, and all of the structures were transformed into α-martensite, as shown in Figure 3d.

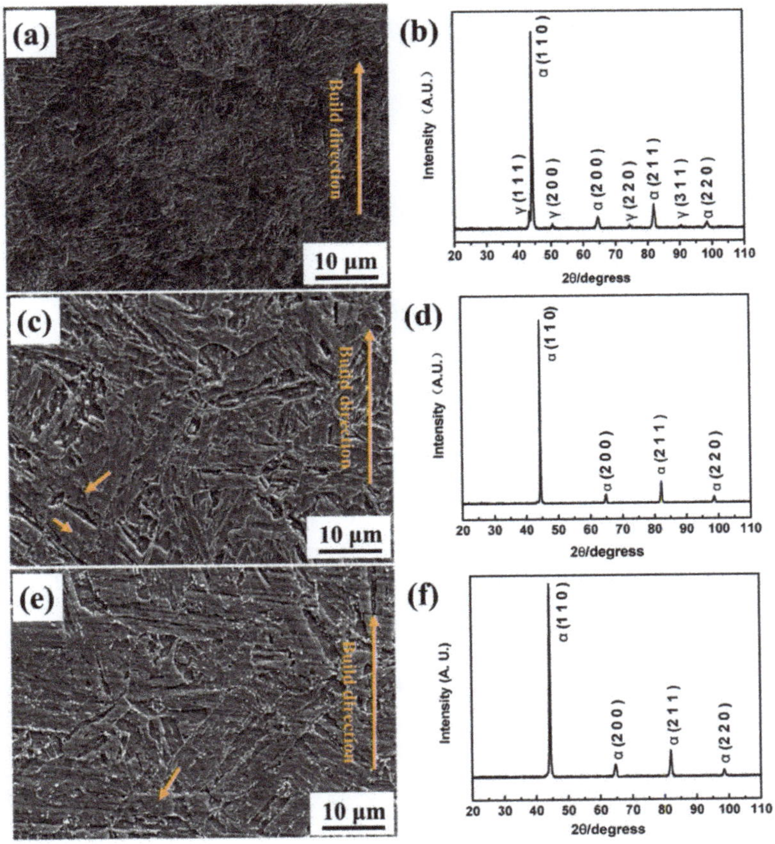

Figure 3. SEM images of the morphology of 17-4PH high-strength steel: (**a**) SLM-processed; (**c**) heat-treated in solution at 940 °C; (**e**) heat-treated in solution at 990 °C. XRD patterns of 17-4PH high-strength steel: (**b**) SLM-processed; (**d**) heat-treated in solution at 940 °C; (**f**) heat-treated in solution at 990 °C.

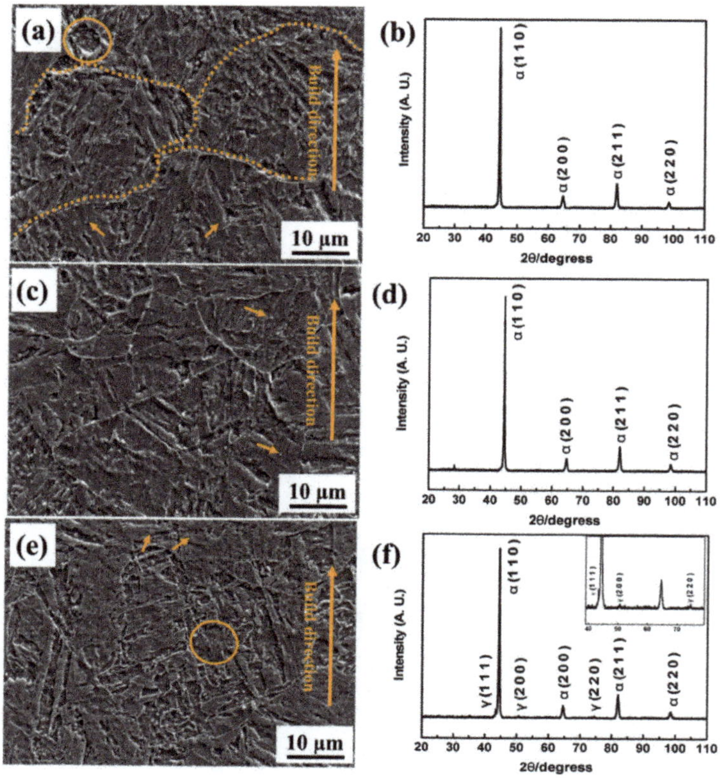

Figure 4. SEM images of the morphology of 17-4PH high-strength steel: (**a**) heat-treated in solution at 1040 °C; (**c**) heat-treated in solution at 1090 °C; (**e**) heat-treated in solution at 1140 °C. XRD patterns of 17-4PH high-strength steel: (**b**) heat-treated in solution at 1040 °C; (**d**) heat-treated in solution at 1090 °C; (**f**) heat-treated in solution at 1140 °C.

The structures were completely austenitized during the heating of the solution and were transformed into lath martensite during the rapid cooling of the solution. When the solution temperature was 1040 °C, as shown in Figure 4a, the original austenite grains were arranged into package shapes (as indicated by the dotted line). Fine equiaxial grains were observed in the structure (the circle mark in the figure). A weak peak of γ-austenite was observed in the spectra of the specimens that were solution-treated at 1140 °C and the XRD patterns are shown in Figure 4f. The main peaks of α-martensite, as well as fine equiaxial grains, were observed (the circle mark in the figure). The starting temperature Ms of the austenite-to-martensite transformation was above room temperature. The solution temperature was too high; thus, Ms decreased and the retained austenite δ-ferrite also increased, but the hardness decreased. At this temperature, the heat treatment effect was not good. The orange arrows in Figures 3 and 4 indicate the precipitated phase. During the aging process, the precipitation of precipitates such as ε-Cu was observed. A high density of dislocations was observed in the martensite, which is conducive to the diffusion of the copper phase. This is likely to improve the strength and hardness of the samples [18].

As shown in Figures 3 and 4, in terms of the XRD patterns, the structure of the 17-4PH high-strength steel samples formed by SLM consisted of the intensive and main peaks of α-martensite and a weak peak of γ-austenite. A complete martensite structure was achieved after the heat treatment. When the solution temperature reached 1140 °C, the characteristic peak of γ (220) appeared again. Owing to the high temperature, Ms decreased, the temperature decreased, some austenite did not transform into martensite, and the amount of residual austenite increased.

3.2. *Effect of the Solution Temperature on the Other Properties*

3.2.1. Microhardness

Figure 5 shows the hardness values of the studied 17-4PH high-strength steel samples that were obtained using the SLM-based method and different solution treatment methods. The hardness improved owing to the heat treatment (the horizontal build direction was the X-Y section, and the vertical build direction was the X-Z section). When the solution temperature was increased to 1040 °C, the X-Y section hardness of the alloy samples peaked at approximately 392 $HV_{0.5}$, while the X-Z section hardness peaked at approximately 381 $HV_{0.5}$. Following that, the hardness of the alloy samples decreased as the solution temperature increased. The increase in the samples' hardness was due to the strengthening effect caused by the complete martensite structure and small precipitates in the samples. The hardness values for the X-Y sections and the X-Z sections of the untreated samples that were made using the SLM-based method were the same. After the heat treatment, the hardness values of the X-Y and X-Z sections were different; the hardness of the X-Z section was lower than that of the X-Y section. For the solution temperature of 940 °C, the hardness was in the 360 $HV_{0.5}$–365 $HV_{0.5}$ range, similar to that of the non-heat-treated samples that were made using the SLM-based method. For the solution temperature of 1140 °C, the hardness of the X-Y section samples decreased to 383 $HV_{0.5}$ and that of the X-Z section decreased to 375 $HV_{0.5}$. This was because, at a solution temperature that was too high, the stability of the austenite increased and the transformation temperature Ms of the martensite decreased, resulting in an uneven martensite structure, a higher amount of retained austenite, higher ferrite content, and lower hardness. After the solution treatment, the microstructure became hard-phase martensite. The dispersed precipitation hardening phase that precipitates on the martensite lath during aging was shown to improve the matrix hardness [22]. Additionally, many dislocations were observed in the structure, which also affected the samples' hardness.

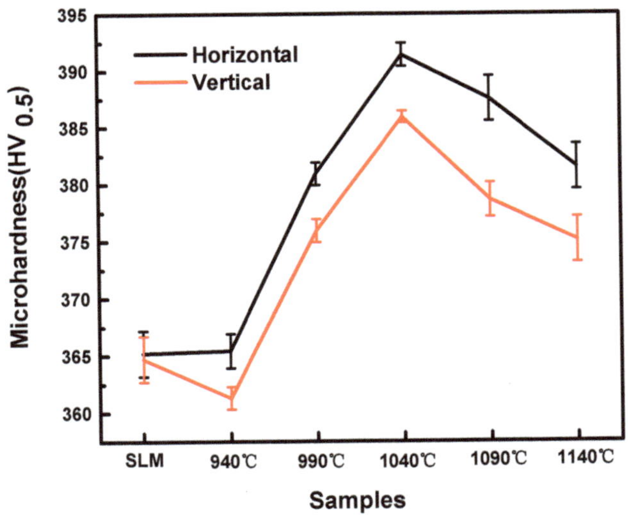

Figure 5. Microhardness of SLM-formed samples and heat-treated samples.

3.2.2. Wear Performance

Figure 6 shows the friction coefficient curves of the 17-4PH high-strength steel samples formed using the SLM-based method and heat-treated at solution temperatures in the 940–1140 °C range. Smooth friction coefficient curves were observed. According to the friction coefficient curves, for all samples, the friction coefficient tended to stabilize after the initial run-in period. The friction coefficient of the untreated samples was approximately 1, while the friction coefficients of the samples obtained for the solution temperatures of 940 °C, 990 °C, and 1140 °C fluctuated strongly around 0.8. The friction coefficient of the sample obtained for the solution temperature of 1040 °C increased rapidly during the first 6 min, and tended to stabilize after 6 min. The average friction coefficient decreased, the friction coefficient was approximately 0.6, and the antifriction performance improved. During friction, the microstructure of the sample's surface was plastically deformed, which increased the actual contact area, resulting in a rapid increase in the friction coefficient. The solution temperature was 1140 °C. Owing to the high solution temperature, the Ms point decreased and the retained austenite δ-ferrite also increased. Austenite is a soft phase; thus, the hardness decreased and the friction coefficient increased.

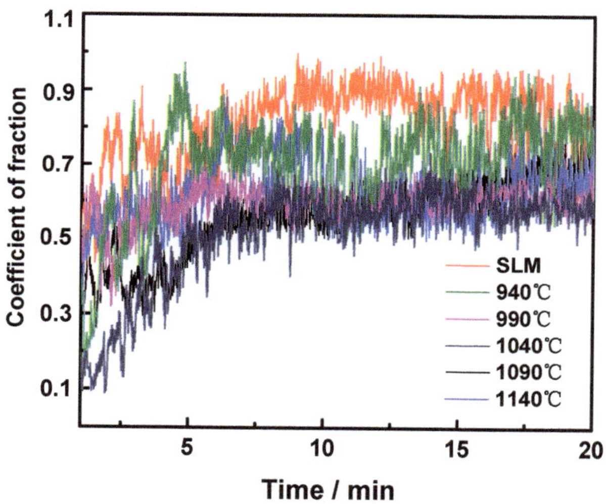

Figure 6. The friction coefficient of SLM-formed samples and samples heat-treated at different solution temperatures.

Figure 7 shows the SEM morphologies of the wear surfaces of the 17-4PH high-strength steel samples obtained using the SLM-based method and heat-treated at different solution temperatures in the 940–1140 °C range. The types of wear that manifested were abrasive wear, adhesive wear, and fatigue wear. During friction, compressive and shear stresses were generated on the samples' surfaces. As shown in Figure 7a, the samples' surfaces became exfoliated, deformed, and scratched during the experiment, which led to the obvious roughening and tearing of the surfaces, as well as to the generation of plate-like wear debris. The material that was peeled off from the specimens' surfaces owing to the propagation of microcracks was the main reason for the observed high wear rates. Figure 7b shows that the adhesive wear for the sample treated at 940 °C was the most obvious, resulting in a large amount of adhesive fine debris. Figure 7c shows that for the solution temperature of 990 °C, the hardness of the corresponding sample increased, and the adhesive wear and lamellar phenomena decreased. Figure 7d shows that for the solution temperature of 1040 °C, the surface of the corresponding sample was the smoothest. Owing to the high hardness and the low degree of wear, the main wear types were abrasive and slight adhesive wear. When the strength of the adhesive joint was lower than that of the two materials engaged in friction, shear occurred at the interface between the two materials. At this time, although the friction coefficient increased, the amount of wear was very small and the material transfer was minimal. Figure 7e shows that, for the solution temperature of 1090 °C, the fatigue wear was clear, and the surface of the sample could be peeled off in flakes, suggesting a high degree of damage. Figure 7f shows that, for the solution temperature of 1140 °C, delamination occurred, yielding a large number of loose wear particles. Some hard particles fell off and wore the surface materials of the contact surface owing to extrusions or relative motion. There were two specific mechanisms of plastic deformation and fracture during the wear process of abrasive particles. When the surface plasticity of the material was high, the abrasive particles scratched, with most of them producing grooves, and accumulation occurred on both sides. Subsequent repeated friction experiments will flatten the accumulated debris. In this way, an external force would eventually cause the formation of cracks and spalling.

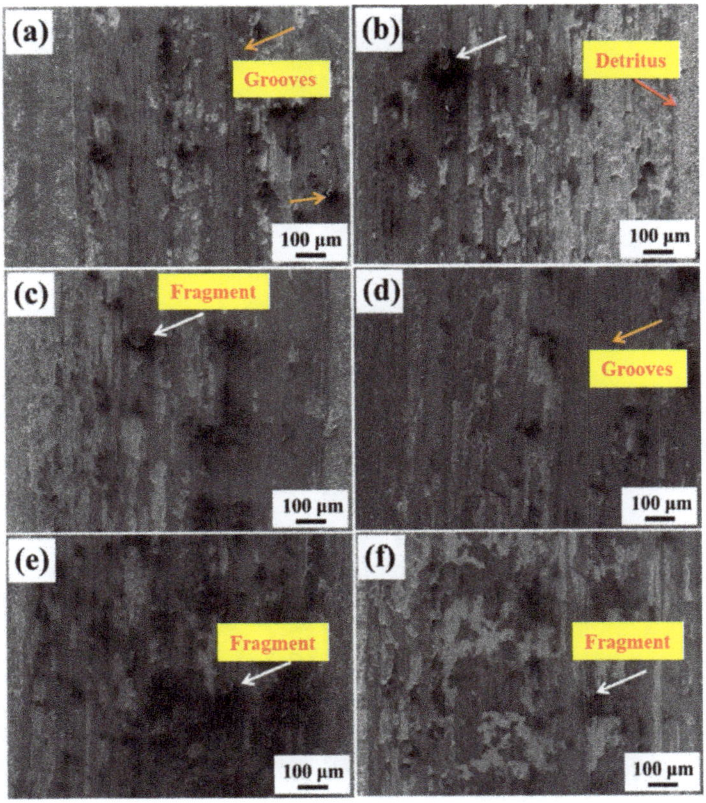

Figure 7. SEM images of the morphology of the wear surface of 17-4PH high-strength steel formed by SLM. (**a**) SLM-formed sample; (**b**) sample heated-treated in solution at 940 °C; (**c**) sample heated-treated in solution at 990 °C; (**d**) sample heated-treated in solution at 1040 °C; (**e**) sample heated-treated in solution at 1090 °C; (**f**) sample heated-treated in solution at 1140 °C.

3.2.3. Tensile Properties

Figure 8 shows the stress–strain curves of 17-4PH high-strength steel formed by SLM and samples heat-treated at different solution temperatures in the 940–1140 °C range. On the whole, the tensile strength of the specimens after heat treatment is improved in different degrees compared with that before heat treatment. Particularly, when the solution temperature is 1040 °C, the trade-off between tensile strength and ductility is better displayed (the tensile strength is 1180 MPa, the strain is 0.16). Compared with Li et al. 's the tensile properties after the homogenizing heat treatment [23], the tensile properties at the solution temperature of 1040 °C are mainly characterized by the fact that the tensile strength is improved to a certain extent compared with that before the heat treatment without sacrificing its toughness. Similarly, compared with Huber et al.'s modified heat treatments (involving prolonged solution annealing and deep-cooling) [24], the trade-off between tensile strength and ductility under the 1040 °C solution heat treatment in this study was also improved.

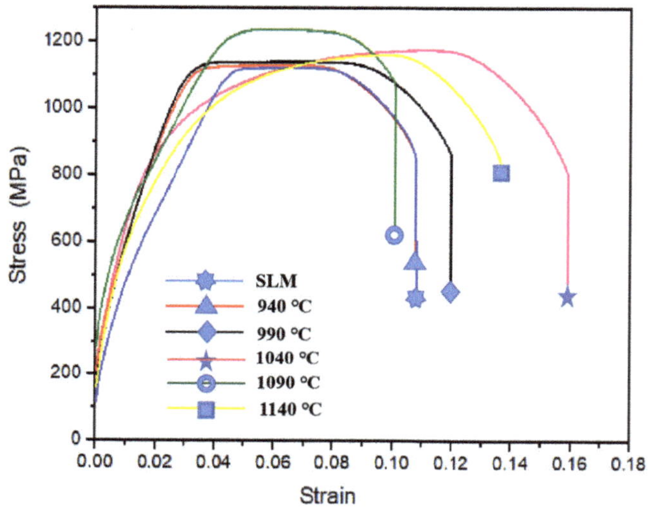

Figure 8. Stress–strain curves of SLM-formed specimens and heat-treated specimens.

Figure 9 shows the SEM morphology results for the tensile fracture of the 17-4PH high-strength steel samples formed using the SLM-based method and processed at different solution temperatures in the 940–1140 °C range. After heat treatment, the tensile strength reached 1200 MPa. The elongation was in the range of 0.10 to 0.16. The tensile fracture surface consists of three distinct regions: the fiber, radiation, and shear lip regions. Figure 9a shows that many dimples and some micropores (orange arrows) formed on the surface of the untreated SLM-formed sample, and the sample exhibited a ductile fracture. As shown in Figure 9b, the sample that was treated at the solution temperature of 940 °C exhibited an obvious cleavage surface (marked by the orange circle) and shallow holes. Dimples were large and uneven, indicating a mixture of ductile and brittle fractures. With an increase in the solution temperature (as shown in Figure 9c for the solution temperature of 990 °C), the torn edge of the tensile part decreased and shortened, and the dimples became finer. As shown in Figure 9d, when the solution temperature increased to 1040 °C, the dimples on the fracture surface of the sample became similar in size and evenly distributed, which was also the reason for the high degree of elongation. The fracture surface of the sample exhibited an obvious necking phenomenon, which is a typical ductile fracture mechanism. As shown in Figure 9e, for the solution temperature of 1090 °C, the tensile strength of the sample increased, but the elongation decreased, resulting in a large number of cleavage surfaces and holes in the sample. As shown in Figure 9f, for the solution temperature of 1140 °C, the austenite in the sample structure reduced the strength of the material but improved the plasticity, tensile rate, cleavage surface, and micropore content of the material. After the heat treatment, the precipitated phase caused the formation of micropores during tension. The speed of the microporous polymerization process affects the toughness of the sample.

The samples' good strength and toughness are related to their structural uniformity and the presence of fine grains.

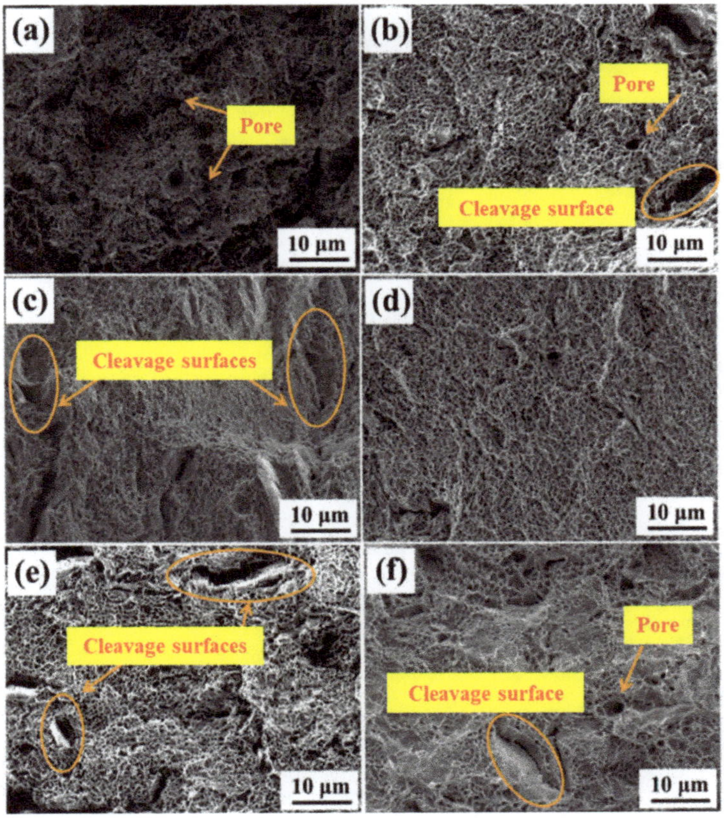

Figure 9. SEM images of the morphology of the tensile fracture of 17-4PH high-strength steel formed by SLM. (**a**) SLM-formed sample; (**b**) sample heat-treated in solution at 940 °C; (**c**) sample heat-treated in solution at 990 °C; (**d**) sample heat-treated in solution at 1040 °C; (**e**) sample heat-treated in solution at 1090 °C; (**f**) sample heat-treated in solution at 1140 °C.

4. Conclusions

In this work, the effects of solution temperature (solution treatment at 940 °C, 990 °C, 1040 °C, 1090 °C, and 1140 °C for 2 h and aging at 480 °C for 4 h) on the microstructure and mechanical properties of 17-4PH high-strength steel samples that were obtained using the SLM-based process. Based on the experimental results, the following conclusions were reached.

First, the 17-4PH high-strength steel is primarily composed of martensite with a small amount of austenite phase and many structural dislocations. After the heat treatment, the martensite grain size increased gradually, and typical martensite microstructures emerged, eventually yielding a single-phase martensite structure. When the solution temperature was increased to 1140 °C, the Ms point decreased and the amount of retained γ-austenite and δ-ferrite also increased, but the hardness decreased. White precipitates were observed in the samples.

Second, the heat treatment improved the hardness, wear properties, tensile strength, and ductility of the 17-4PH high-strength steel samples formed using the SLM-based process. After the solution treatment at 1040 °C for 2 h and aging at 480 °C for 4 h, the

Vickers hardness of the 17-4PH high-strength steel samples increased to 392 $HV_{0.5}$, with a 7% improvement, and the friction coefficient was approximately 0.6, which was 5% higher than that of the untreated samples.

Author Contributions: Writing—original draft, Z.Z.; writing—original draft, investigation, data analysis and data collection, H.W.; writing—review and editing, P.H.; supervision, P.B.; conceptualization and methodology, W.D.; study design, X.L.; supervision, J.L.; conceptualization and methodology, W.Z. All authors have read and agreed to the published version of the manuscript.

Funding: This research was funded by the Major Science and Technology Projects of Shanxi Province, China (No. 20181101009), the Key Research and Development Project of Shanxi Province (201903D121009), and the Scientific and Technological Innovation Projects of Shanxi Province, China (2019L0608).

Institutional Review Board Statement: Not applicable.

Informed Consent Statement: Not applicable.

Data Availability Statement: Data are contained within the article.

Acknowledgments: The authors would like to thank the Major Science and Technology Projects of Shanxi Province, China (No. 20181101009), the Key Research and Development Project of Shanxi Province (201903D121009), and the Scientific and Technological Innovation Projects of Shanxi Province, China (2019L0608).

Conflicts of Interest: The authors declare no conflict of interest.

References

1. Bressan, J.D.; Daros, D.P.; Sokolowski, A.; Mesquita, R.A.; Barbosa, C.A. Influence of hardness on the wear resistance of 17-4 PH stainless steel evaluated by the pin-on-disc testing. *J. Mater. Process. Technol.* **2008**, *205*, 353–359. [CrossRef]
2. Dong, H.; Esfandiari, M.; Li, X.Y. On the microstructure and phase identification of plasma nitrided 17-4PH precipitation hardening stainless steel. *Surf. Coat. Technol.* **2008**, *202*, 2969–2975. [CrossRef]
3. Alnajjar, M.; Christien, F.; Barnier, V.; Bosch, C.; Wolski, K.; Fortes, A.D.; Telling, M. Influence of microstructure and manganese sulfides on corrosion resistance of selective laser melted 17-4 PH stainless steel in acidic chloride medium. *Corros. Sci.* **2020**, *168*, 108585. [CrossRef]
4. Murr, L.E.; Martine, E.; Hernandez, J.; Collins, S.; Amato, K.N.; Gaytan, S.M.; Shindo, P.W. Microstructures and Properties of 17-4 PH Stainless Steel Fabricated by Selective Laser Melting. *J. Mater. Res. Technol.* **2012**, *1*, 167–177. [CrossRef]
5. Lin, X.; Cao, Y.Q.; Wu, X.Y.; Yang, H.O.; Cheng, J.; Huang, W.D. Microstructure and mechanical properties of laser forming repaired 17-4PH stainless steel. *Mater. Sci. Eng. A* **2012**, *553*, 80–88. [CrossRef]
6. Hsiao, C.N.; Chiou, C.S.; Yang, J.R. Aging reactions in a 17-4 PH stainless steel. Mater. *Chem. Phys.* **2002**, *74*, 134–142.
7. Yoo, W.D.; Lee, J.H.; Youm, K.T.; Rhyim, Y.M. Study on the Microstructure and Mechanical Properties of 17-4 PH Stainless Steel Depending on Heat Treatment and Aging Time. *Solid State Phenom.* **2006**, *118*, 15–20. [CrossRef]
8. Yadollahi, A.; Shamsaei, N. Additive manufacturing of fatigue resistant materials: Challenges and opportunities. *Int. J. Fatigue* **2017**, *98*, 14–31. [CrossRef]
9. Holland, S.; Wang, X.Q.; Fang, X.Y.; Guo, Y.B.; Yan, F.; Li, L. Grain boundary network evolution in Inconel 718 from selective laser melting to heat treatment. *Mater. Sci. Eng. A* **2018**, *725*, 406–418. [CrossRef]
10. Ma, M.; Wang, Z.; Zeng, X. A comparison on metallurgical behaviors of 316L stainless steel by selective laser melting and laser cladding deposition. *Mater. Sci. Eng. A* **2017**, *685*, 265–273. [CrossRef]
11. Sun, Y.; Hebert, R.J.; Aindow, M. Effect of Laser Scan Length on the Microstructure of Additively Manufactured 17-4PH Stainless Steel Thin-Walled Parts. *Addit. Manuf.* **2020**, *35*, 101302. [CrossRef]
12. Bajguirani, H.H.; Servant, C.; Cizeron, G. TEM investigation of precipitation phenomena occurring in PH 15-5 alloy. *Acta Metall. Mater.* **1993**, *41*, 1613–1623. [CrossRef]
13. Bhambroo, R.; Roychowdhury, S.; Kain, V.; Raja, V.S. Effect of reverted austenite on mechanical properties of precipitation hardenable 17-4 stainlesssteel. *Mater. Sci. Eng. A* **2013**, *568*, 127–133. [CrossRef]
14. Si, Q.; Qin, F.; Li, K.F.; Liu, X.; Zhou, G. Effect of hot isostatic pressing on the microstructure and mechanical properties of 17-4PH stainless steel parts fabricated by selective laser melting. *Mater. Sci. Eng. A* **2021**, *810*, 141035.
15. Yadollahi, A.; Mahmoudi, M.; Elwany, A.; Doude, H.; Newman, J.C. Effects of Crack Orientation and Heat Treatment on Fatigue-Crack-Growth Behavior of AM 17-4 PH Stainless Steel. *Eng. Fract. Mech.* **2020**, *226*, 106874. [CrossRef]
16. Sun, Y.; Hebert, R.J.; Aindow, M. Effect of heat treatments on microstructural evolution of additively manufactured and wrought 17-4PH stainless steel. *Mater. Des.* **2018**, *156*, 429–440. [CrossRef]
17. Cheruvathur, S.; Lass, E.A.; Campbell, C.E. Additive Manufacturing of 17-4 PH Stainless Steel: Post-processing Heat Treatment to Achieve Uniform Reproducible Microstructure. *JOM* **2016**, *68*, 930–942. [CrossRef]

18. Chung, C.Y.; Tzeng, Y.C. Effects of aging treatment on the precipitation behavior of ε-Cu phase and mechanical properties of metal injection molding 17-4PH stainless steel. *Mater. Lett.* **2019**, *237*, 228–231. [CrossRef]
19. Barroux, A.; Ducommun, N.; Nivet, E.; Laffont, L.; Blanc, C. Pitting corrosion of 17-4PH stainless steel manufactured by laser beam melting. *Corros. Sci.* **2020**, *169*, 108594. [CrossRef]
20. Lou, X.Y.; Andresen, P.L.; Rebak, R.B. Oxide inclusions in laser additive manufactured stainless steel and their effects on impact toughness and stress corrosion cracking behavior. *J. Nucl. Mater.* **2018**, *499*, 182–190. [CrossRef]
21. Kong, D.; Ni, X.; Dong, C.; Zhang, L.; Man, C.; Yao, J.; Xiao, K.; Li, X. Heat treatment effect on the microstructure and corrosion behavior of 316L stainless steel fabricated by selective laser melting for proton exchange membrane fuel cells. *Electrochim. Acta* **2018**, *276*, 293–303. [CrossRef]
22. Yeli, G.; Auger, M.; Wilford, K.; Smith, G.; Bagot, P.; Moody, M. Sequential nucleation of phases in a 17-4PH steel: Microstructural characterisation and mechanical properties. *Acta Mater.* **2017**, *125*, 38–49. [CrossRef]
23. Li, K.; Sridar, S.; Tan, S.; Wei, X. Effect of homogenization on precipitation behavior and strengthening of 17-4PH stainless steel fabricated using laser powder bed fusion. *arXiv* **2012**, arXiv:2112.06289.
24. Huber, D.; Stich, P.; Fischer, A. Heat Treatment of 17–4 PH Stainless Steel Produced by Binder Jet Additive Manufacturing (BJAM) from N_2-Atomized Powder. *Prog. Addit. Manuf.* **2021**. [CrossRef]

MDPI
St. Alban-Anlage 66
4052 Basel
Switzerland
Tel. +41 61 683 77 34
Fax +41 61 302 89 18
www.mdpi.com

Metals Editorial Office
E-mail: metals@mdpi.com
www.mdpi.com/journal/metals

www.ingramcontent.com/pod-product-compliance
Lightning Source LLC
Chambersburg PA
CBHW061622210926
51850CB00009B/2404
* 9 7 8 3 0 3 6 5 6 8 6 5 2 *